RECENT TRENDS IN BIOTECHNOLOGY

SOLID-STATE CULTIVATION OF *GRIFOLA FRONDOSA* AND *GANODERMA LUCIDUM*

TECHNOLOGY, RESEARCH AND POSSIBILITIES FOR WASTE UTILIZATION IN TROPICAL WEATHERS

RECENT TRENDS IN BIOTECHNOLOGY

Additional books and e-books in this series can be found
on Nova's website under the Series tab.

RECENT TRENDS IN BIOTECHNOLOGY

SOLID-STATE CULTIVATION OF *GRIFOLA FRONDOSA* AND *GANODERMA LUCIDUM*

TECHNOLOGY, RESEARCH AND POSSIBILITIES FOR WASTE UTILIZATION IN TROPICAL WEATHERS

SANDRA MONTOYA
ÓSCAR J. SANCHEZ
AND
DIEGO F. ARIAS

nova
science publishers
New York

Library of Congress Cataloging-in-Publication Data

ISBN: 978-1-53616-481-7
Library of Congress Control Number:2019951534

Published by Nova Science Publishers, Inc. † New York

CONTENTS

Preface

At present, the planet is facing several situations that threaten the survival of the human species. The amount of waste generated daily that has marginal uses, the shortage of food, and the emergence of new diseases make it necessary to seek new and sustainable solutions to these problems. Mankind should rely on research and development to address these serious issues. To this end, the conversion of waste into feedstocks for different production processes, the development of new products for pharmacological use, the creation of new technological processes for the bioremediation and treatment of waste that are difficult to degrade, and the return of lignocellulosic biomass to productive farms as fertilizers or animal feed, among other alternatives, are of paramount importance for humanity. For these reasons, fungi are wonderful organisms that offer important tools for facing these challenges.

Fungi are heterotrophic organisms that lack chlorophyll. They can be unicellular organisms, like yeasts, and they can also be multicellular, like filamentous fungi. They engage in sexual and asexual reproduction. Unicellular organisms are microtubular machineries that can only be divided between daughter cells, while multicellular organisms contain cells that are specialized for cell division and other related functions. Basidiomycetes is a class of fungi represented by more than 30,000 species, the metabolism and production capacity of which has not been investigated in depth.

Basidiomycetes produce a wide range of natural products, ranging from structural components with antitumor and immunomodulatory activities to antimicrobial, antifungal, antiviral, cytostatic, enzymatic, growth-regulating, and flavoring activities, among others. These fungi play an important role in the efficient biodegradation of lignocellulosic materials and other natural biopolymers as well as in the degradation of materials that are difficult to degrade, such as xenobiotics. A special group of fungi, the white rot basidiomycetes, can be used for producing valuable biologically active substances. Among these substances, lignocellulolytic enzymes (mainly hydrolases and oxidoredu-ctases) should be highlighted in consideration of their different industrial applications and their potential use in the synthesis of specialty chemicals. In particular, the use of lignocellulosic residues as raw materials for substrate preparation during the cultivation of white rot fungi represents a very important scientific and technological option for obtaining a wide range of biologically active metabolites as well as the production of useful fruiting bodies for the food and pharmaceutical industries.

Ganoderma lucidum and *Grifola frondosa* are wood-degrading fungi, and they cause white rot. These fungi both exhibit multiple medicinal attributes, and they differ in their textures and edibility. *G. frondosa* (known as maitake) may be used for these two purposes, while *G. lucidum* (known as reishi) is used exclusively for medicinal purposes. Currently, these two fungal species are being studied by groups around the world. However, more research has been performed on *G. lucidum*. These studies are related to its biology and artificial cultivation, the production of its metabolites by solid-state and submerged fermentation, the production of purified substances or extracts, and tests of its biological activity against cancer, cardiovascular diseases, diabetes, and others. In contrast, the research on *G. frondosa* is much more recent but no less important. Maitake is an edible fungus with nutritional attributes and multiple medicinal properties that have been proven by many researchers; applications of maitake include use as a coadjuvant for chemotherapy and radiotherapy treatments, an immune system enhancer, and a cellular regenerator, among others.

A substantial quantity of scientific and technical documents (research papers, reviews, books, book chapters, and patents) related to the artificial cultivation of reishi and maitake and their bioactive substances have been published. However, most of these studies have been performed in North American, European, and Asian countries, and there have been fewer studies in Latin American countries. In particular, numerous published scientific articles and books have presented different aspects of the general mycology, genetics, physiology, and fungal biochemistry of these two fungal species with different degrees of detail. In addition, some texts and articles dedicated to the cultivation technologies used for these mushrooms are available. Nevertheless, as far as the authors know, there are no publications on the cultivation specifications for *G. lucidum* and *G. frondosa* on different production scales in tropical climates, and none of these publications is accompanied by a general overview of the raw materials available for substrate preparation in tropical countries. Moreover, there is little information on the specific bioactive substances produced by these two species, such as the lignocellulolytic enzymes secreted by these two white rot fungi or the degradation of the lignocellulosic materials that they use as substrates.

Therefore, we present this book, entitled "Solid-state Cultivation of *Grifola frondosa* and *Ganoderma lucidum*: Technology, Research and Possibilities for Waste Utilization in Tropical Climates" to help close this knowledge gap, albeit partially. Our objective is to offer scientists, researchers, engineers, undergraduate and postgraduate students, the technical staff of mushroom production companies, and people interested in reishi and maitake fungi information on the cultivation parameters for producing these two mushrooms, especially their fruiting bodies, with high biological efficiency under tropical weather conditions. We also address the basic foundations of the relevant biology and conservation procedures, among other related aspects, including the production of bioactive compounds and lignocellulolytic enzymes.

All the information disclosed in this book is based on the research activities conducted by the authors for more than twelve years at the Universidad de Caldas, located in the municipalities of Manizales and

Villamaría (Colombia). In particular, most of the results presented in this book were obtained during the project entitled *Implementación de una estrategia integral a través de innovación biotecnológica para el aprovechamiento de residuos en el Departamento de Caldas* [Implementation of an integral strategy through biotechnological innovation for waste utilization in the Department of Caldas], which was funded by the *Fondo de Ciencia, Tecnología e Innovación del Sistema General de Regalías de Colombia* [Fund of Science, Technology, and Innovation of the Colombian General Royalties System, grant number 2012000100178] and the Universidad de Caldas [grant 0700413]. In addition, the scientific and commercial information on the cultivation of maitake mushrooms corresponds to the outcomes obtained during the execution of a project entitled *Desarrollo y evaluación de los parámetros de cultivo a escala semi-industrial del hongo Grifola frondosa en climas tropicales y su transformación con fines comerciales como productos nutracéuticos* [Development and evaluation of the parameters of semi-industrial scale cultivation of the fungus *Grifola frondosa* in tropical climates and its transformation for commercial purposes as nutraceutical products] within the framework of Call 523-2011, "Project Bank for Creation of Technology-based Companies or Business Units" and was funded by the *Departamento Administrativo de Ciencia, Tecnología e Innovación de Colombia, Colciencias* [Colombian Administrative Department of Science, Technology and Innovation, Colciencias]. It is worth highlighting that the first project conducted by the authors on maitake, which was entitled *Selección, desarrollo y estabilización de un inóculo del macromiceto Grifola frondosa* [Selection, development, and stabilization of an inoculum of the *Grifola frondosa* macromycete] was funded by the Universidad de Caldas and SENA Regional Caldas (started in 2006), and the project *Evaluación de la producción de Ganoderma lucidum y su actividad antiinflamatoria bajo dos condiciones de cultivo en estado sólido: luz azul y suplementación con germanio 132* [Evaluation of *Ganoderma lucidum* production and its anti-inflammatory activity under two solid-state cultivation conditions: Blue light and supplementation with germanium 132] was funded by Universidad de Caldas and the Universidad Nacional de

Colombia sede Manizales within the framework of Call 2015, "Stamp pro University of Caldas and National University of Colombia, Manizales".

This book is divided into eight chapters. The first three chapters present *G. lucidum* research results, such as the basic biology of this mushroom, its market, the selection of strains and their conservation, feedstocks for its production, solid-state cultivation at different scales, and the primary research trends in reishi cultivation and its biologically active compounds. In Chapters 4-7, research results on *G. frondosa* related to the basic biology of this mushroom, its market, the selection of strains and their conservation, feedstocks for its production, solid-state cultivation at different scales (including its industrial production), and the primary research trends in Reishi cultivation and its biologically active compounds are presented. The final chapter covers basic concepts about the lignocellulolytic enzymes produced by white rot fungi, as well as some research on the production of these enzymes by reishi and maitake mushrooms, including the description and application of a novel solid-state bioreactor for producing bioactive substances that was designed, built, and put in motion by the authors.

The material presented in this book includes basic concepts relating to biology, microbiology, and enzymology, as well as information published by researchers from around the world on reishi and maitake. In addition, the primary focus of this monograph is that the results obtained by the authors are presented in detail through 30 research cases. These cases briefly describe the nonpublished data obtained during the execution of the abovementioned projects as well as outcomes published in previous studies and derived from PhD and MSc theses. Almost all the information in these cases was generated through numerous experiments performed at the Bioprocess and Agro-industry Plant, a center for research and technological development at the Universidad de Caldas, for which the main goal is to provide scientific knowledge and technical information for technology transfer to the Colombian industrial sector and communities as well as to support the education of undergraduate and postgraduate students. Currently, Doctor Sandra Montoya is Director of the Bioprocess and Agro-industry Plant and is an active investigator at the Research Group on Food and Agro-industry in the Faculty of Engineering at the Universidad de

Caldas. Doctor Óscar J. Sánchez is Head of the Research Group on Food and Agro-industry and is a researcher associated with the Bioprocess and Agro-industry Plant, and Agronomist Diego F. Arias is Supervisor at the Bioprocess and Agro-industry Plant.

The authors wish to thank Doctors Laura Levin, Maryeimy Varón, Carlos E. Orrego, and Belarmino Segura and Engineer Diana M. López, who helped the authors during the *G. frondosa* investigation. In addition, the authors acknowledge the administrative support of Patricia Salazar, Vice-Chancellor of University Projection, for assisting during the execution of the project funded by the General Royalties System as well as the Vice-Rectorate on Research and Postgraduate Programs at the Universidad de Caldas. In addition, we also express our thanks to our postgraduate students, administrative staff, and colleagues who helped us in one way or another during the research projects on which this book was based.

Chapter 1

MARKET AND FEEDSTOCKS FOR *GANODERMA LUCIDUM* CULTIVATION

INTRODUCTION

Ganoderma lucidum (Curt.: Fr) P. Karst (lingzi in Chinese; reishi, mannentake, or sachitake in Japanese; and youngzi in Korean) is a species in the class Basidiomycetes belonging to the Polyporaceae (or Ganodermataceae) family of the Aphylloporales order. *G. lucidum* is a fungus that degrades wood and causes white rot (Chang and Miles 2004). The fruiting bodies of *G. lucidum* have become more popular over the last two decades due to their health-improving properties. Currently, dietary supplements prepared from the *G. lucidum* mycelium or carpophores are found in different markets all around the world.

Considering the medicinal attributes of this fungus, research has been extended to the exploration of other *Ganoderma* species. More than 250 species have been described and studied, including *G. australe* (Fr) Pat., *G. applanatum* (Pers. Ex Wallr) Pat., and *G. tsugae* Murr., Bull, among others. However, the valuable medicinal attributes continue to be reported in *G. lucidum* strains when evaluating the therapeutic practices and scientific literature reported to date (Chen and Yu 1991, Kim and Kim 2002).

G. lucidum is a species of white rot fungi, i.e., fungi that are capable of degrading the lignin component of lignocellulosic substrates other than the cellulose component. White rot fungi are so called because the degradation process results in the bleaching of the wood substrate. During this process, the lignin undergoes oxidative degradation due to the action of the ligninases synthesized by these fungi. In addition, white rot fungi secrete a large amount of cellulases to hydrolyze the cellulose contained in the lignocellulosic materials.In general, fungi are presented as organisms that have unlimited potential in the recycling of organic matter. The development of white rot fungi requires culture media (substrates) composed of materials that will ensure the physical characteristics and supply the chemical components suitable for their growth and development under controlled conditions. Lignocellulosic materials meet the general conditions required by these fungi.

In fact, the lignocellulosic complex composed of cellulose, hemicellulose, and lignin is the most abundant biopolymer on the planet, as it is the main component of plants and their residues as well as of agro-industrial, forestry, and urban solid waste, among others (Sánchez and Cardona 2007). The main biopolymers in the plants are starch (which is used as an energy reserve) and those that make up their cell walls: lignin, cellulose, hemicellulose, and pectin. In wood, the cell walls are highly lignified, and the pectin content is very low or zero (Papinutti, Dorio, and Forchiassin 2003).In this chapter, the general characteristics of the reishi mushroom market are presented. In addition, an overview on the different raw materials employed for producing *G. lucidum* is provided. Finally, the main chemical components of these materials are explored.

General Aspects of the Reishi Mushroom Market

Global Medicinal Mushroom Market

It has been observed that maintaining a nutritional balance and an active lifestyle in a suitable environment can help people to achieve optimal health throughout life. Mushrooms have been used by ordinary people and the imperial court of China for 2,000 years as a dietary therapy to maintain or improve health or to treat diseases. For Chinese society, "food and medicines have a common origin", which makes fungi an example of this ancient Chinese saying. This statement is particularly applicable to mushrooms, whose nutritional qualities and tonic effects as nutriceuticals (Chang and Buswell 1996) or as dietary supplements have been recognized for a long time (Wasser 2010). Chang and Wasser (2012) explained that human health can be divided into three states: health, secondary health, and disease. Thus, mushrooms can be used primarily as food to support a healthy state, as a medicine for diseases, and as a dietary supplement for a sub-healthy state or secondary health.

The global mushroom market was estimated at more than $35.08 billion (US dollars) in 2015 according to Zion Market Research. The estimated global production of mushrooms and truffles reached 10,790,859 tonnes in 2016 according to FAO statistics (FAO 2018), with eastern Asia being the region with the highest share (73.2%) and China being the main producer (7,786,399 tonnes). The mushroom market is anticipated to reach more than USD 59.48 billion in 2021 and is expected to increase at a compound annual growth rate of slightly more than 9.2% from 2016 to 2021 (Zion Market Research 2018). The Asia-Pacific region is expected to show the highest growth in the mushroom market owing to escalating demand from the food industry. Mushroom consumption is high in China, Japan, and India due to its high nutritional content. The increasing prevalence of various diseases due to hectic lifestyles has led many people to turn towards organic and healthy food. This interest is likely to increase the demand for mushrooms in this region. Likewise, the rapidly expanding food industries in India and China will positively impact the mushroom market (CNBC 2018, Zion Market Research 2018). Europe is the largest market for the cultivation and processing of mushrooms. Europe accounted for more than 35% of the total share of the mushroom processing and cultivation market. Notably, there has been a tremendous growth in demand for organic food in North America due to the growing incidence of diabetics and obesity in this region. In this context, mushrooms have valuable nutritional properties, such as low cholesterol and no fat. Hence, mushrooms are being used as a staple food for patients suffering from obesity, high cholesterol, and other conditions. Because there has been a shift in preferences for healthy and organic food among the health-conscious population, the growth of the mushroom market in North America and Europe has grown significantly.

Latin America has been witnessing an unprecedented boom in recent times due to its high use of mushroom as ingredients in various food products. In particular, the Brazilian market offers several processed food products due to high demand from consumers. This trend is expected to trigger growth in the mushroom market. The Middle East and Africa will exhibit notable growth in the coming years due to the abundance of raw resources and favorable conditions for mushroom growth (CNBC 2018, Zion Market Research 2018).

Nevertheless, the short shelf life of mushrooms increases their costs to market players. Although this disadvantage might deaccelerate the growth of the mushroom market, growing food markets and food chains in developing nations along with ongoing research on the untapped applications of mushrooms is expected to unlock new ways to develop the mushroom market in the coming years (Zion Market Research 2018). The global mushroom market is segmented by application as follows: Fresh mushrooms, processed mushrooms, dried mushrooms, canned mushrooms, frozen mushrooms, and others. To the end user, this market is oriented towards the food industry, medical uses, and direct consumption. The main mushrooms that are commercialized are as follows: milky mushrooms, button mushrooms, shiitake mushrooms, oyster mushrooms, paddy mushrooms, winter mushrooms, and reishi mushrooms.

Many mushrooms, such as *G. lucidum*, *Coriolus versicolor*, *Hericium erinaceus*, *Armillaria mellea*, *Marasmius androsaceus*, *Tremella fuciformis*, and *Lentinus edodes*, have traditionally been used as medicines and tonics in China, Korea, and Japan. Terpenoids, substances that can inhibit the growth of cancer cells, have been found in *G. lucidum*. Steroids in *C. versicolor* and *Agaricus blazei*, the novel γ-pyrones esinapyrones A and B, the novel cytotoxic phenols hericenones A and B, and a novel fatty acid in *Hericium erinaceus* as well as lampterol in *Lampteromyces japonicus* represent other inhibitors of cancer cell growth. The germanium contained in mushrooms is involved in carcinostatic and analgesic functions. Many polysaccharides that exhibit remarkable antitumor activity in vivo have been isolated from various mushroom species. These antitumor substances are regarded as biological response modifiers, i.e., immunotherapeutic agents, and their activities are based on the activation of immunological functions in the host (Rajarathnam and Shashirekha 2003).

The medicinal attributes of mushrooms have allowed for the rapid growth of the mushroom nutraceuticals market. These nutraceuticals are medicinal preparations from mushrooms, and they are functional foods that are enriched or modified and consumed as part of a normal diet to provide health-giving benefits. They are refined or partially refined extracts consumed in the form of capsules or tablets, as a dietary supplement (not a food), and they have potential therapeutic applications. Nutraceuticals are not pharmaceuticals, which are defined as chemical preparations, and they are mainly used therapeutically. Because pharmaceuticals have high potency, the regulations by the authorities are required to prevent the chance of overdose. Mushrooms with little possible of toxicity and overdose do not require very strict regulation and can be sold as 'over-the-counter' medicines (Rajarathnam and Shashirekha 2003). In the recent past, a variety of medicinal preparations in the forms of tablets, capsules, and extracts from mushrooms have been produced and marketed.

Since the end of World War II, the industrial-scale production of mushrooms with edible and medicinal attributes has increased steadily. The global production of edible fungi cultivated from the 1960s to the year 2012 is shown in Table 1. From 1960 to 1970, there was an increase in the production of edible fungi of greater than 500%. For 1990, the total production was 3.79 million tonnes, increasing to 6.16 million tonnes in 1997. In general, the global production of mushrooms increased more than 12% annually from 1981 to 1997, with a 31% increase in global production between 2009 and 2012. However, mushrooms in the *Agaricus* genus decreased as a percentage of total global production, most likely because alternative edible mushrooms now have higher demand. For example, the percentage of total global mushroom consumption corresponding to

Lentinus edodes (known as shiitake) increased from 14.3% in 1981 to 25.2% in 1997, and its total global production increased from 180 thousand tonnes to 1,564.4 million tonnes. In addition, the fungi *Pleurotus* spp. showed increased global consumption, from 2.8% to 14.2%, and their production grew from 35 thousand tonnes to 875.6 thousand tonnes, with a 25-time increase within the same period of time. It is necessary to highlight that for the year 2012, China produced more than 85% of the edible mushrooms available worldwide (Chang and Wasser 2017).

Table 1. Global production of cultivated edible mushrooms, fresh equivalent, from 1960–2009

Year	Production (× 1,000 tonnes)	Increase (%)	Annual Increase (%)
1960	170.0	-	-
1965	341.0	100.6	20.1
1970	546.0	60.0	12.0
1975	916.0	67.8	13.6
1981	1,257.2	37.3	6.2
1986	2,176.0	73.1	14.6
1990	3,794.0	74.4	18.6
1991	4,273.0	12.6	12.6
1994	4,909.3	14.9	5.0
1997	6,158.4	25.4	8.5
2002	12,250.0	98.9	19.8
2009	24,000.0	95.9	13.7
2012	31,500.0	31.3	10.4

Source: Chang and Wasser (2017).

Reishi Market

Reishi is a medicinal fungus with a long history of use as a Chinese folk medicine for the promotion of health and longevity. It is also widely used as a traditional medicine in Korea and Japan. In general, this mushroom has been used traditionally in these regions for its health promoting characteristics, such as enhanced immunity, longevity, liver and heart health promotion and stress relief because it acts as a slight central nervous system sedative, which relaxes the mind and body (Persistence Market Research 2018). Since ancient times in China, *G. lucidum* has been used to improve health and treat diseases by the broader society, in general, and by the rulers of the empire, in particular. *G. lucidum* can be used in the diet to maintain health as a health enhancer, as a dietary supplement or as a tonic (nutraceutical) to treat diseases (Chang and Wasser 2017). For hundreds of years, the medicinal value of Lingzhi (or Reishi) was treasured in China because this fungus is rarely found in nature. This lack of availability was largely responsible for this fungus being so precious and expensive. During ancient times, anyone who picked this mushroom from the natural environment and presented it to a high-ranking official in China was generally well rewarded. The reishi is a focal point in ancient Chinese and Japanese artwork and has been associated with royalty, wisdom, sexual prowess, and eternal life (Stamets 2000). References to reishi as a superior herb that enhances human health can be found as early as 100 B.C. Currently, *G. lucidum* species continue to be prescribed in traditional medicine, for which its fruiting bodies are typically grown, ground, and made into pills, tinctures, or teas (Loyd et al. 2018). At present, products based on *G. lucidum* have been considered as high value products.

Modern medicinal studies indicate that *G. lucidum* is effective at preventing or relieving various human diseases. Its medicinal properties include anti-aging effects, blood pressure lowering, and immunity enhancement, as well as the prevention and treatment various cancers, chronic bronchitis, diabetes mellitus, gastric ulcers, hepatitis, hyperlipidemia, hypertension, neurasthenia and thrombosis. This fungus is also used in various cosmetics (Cao, Wu, and Dai 2012). Its medicinal effects come from its metabolites, including polysaccharides, triterpenes, lucidenic acids, LZ-8 protein, adenosine, ergosterol, glucosamine, and cerebrosides. Numerous in vitro and animal studies have demonstrated the antitumor and immunomodulatory effects of reishi mushrooms. Romm et al. (2010) have summarized these reports about the primary medicinal properties of *G. lucidum* as follows. A wide range of antitumor and immunomodulatory functions have been purported and observed when using the water extract and the polysaccharide fraction as well as the alcohol extract or the triterpene fraction, and these functions include the enhanced activity of antigen-presenting cells, the mononuclear phagocyte system, humoral immunity, and cellular immunity. The reishi polysaccha-ride peptide (Gl-PP) has demonstrated antitumor effects in mice and potential antiangiogenesis; therefore, inducing cell apoptosis might be one of the mechanisms of action that inhibits human carcinoma cells. It has been hypothesized that the key attribute underlying the antiangiogenic potential of Gl-PP is that it may directly inhibit vascular endothelial cell proliferation or indirectly decrease the growth factor expression of tumor cells. It has been demonstrated that *G. lucidum* induces apoptosis, inhibits cell proliferation, and suppresses the cell migration of highly invasive human prostate cancer cells. Experimental results evaluating humoral immunity in horses have shown that *G. lucidum* could help horses to produce a significantly higher quantity of specific antibodies within a shorter time. *G. lucidum* has also shown antioxidant, free radical scavenging, and chelating abilities. In addition, *G. lucidum* produces the antifungal protein ganodermin, which has inhibitory effects against common fungi, such as *Botrytis cinerea* and *Fusarium oxysporum*, and it also produces other chemicals with antibacterial and nematacidal effects (Loyd et al. 2018).

Reishi Mushroom Extract

G. lucidum is currently among the most popular medicinal mushrooms on the global market. Several reishi products made from the cultivated fruiting bodies have been commercialized as dietary supplements around the world, especially in Asian countries; its estimated annual global turnover is approximately $2.16 billion (Cao, Wu, and Dai 2012). The dietary supplement industry, which consists of vitamins, minerals, botanicals, etc., is a growing market with global sales of $109 billion and an expectation to be nearly double that by 2020. Based on these figures, the reishi industry accounts for approximately 2% of global dietary supplement sales (Loyd et al. 2018).

The extracts obtained from *G. lucidum* are composed of complex carbohydrates or water-soluble polysaccharides, triterpenoids, proteins, and amino acids. These extracts have antioxidant, cholesterol-lowering, anticancer, and immunological properties. Thus, they are used for treating diabetes, respiratory problems, viral infections, heart and kidney diseases, liver problems, insomnia, anxiety, nervous system disorders, and others. For these reasons, the global reishi mushroom extract market is mainly driven by rapidly growing demand and increasing popularity in the healthcare industry. Moreover, increasing the use of this extract in the developing personal care and food and beverage industries has supplemented the market growth (Allied Market Research 2018). This increase has resulted in the inclusion of this extract in the product portfolio of the key players. Furthermore, because of its health benefits, the product has witnessed increased traction, particularly in developed markets. All these factors have stimulated the growth of the global reishi mushroom extract market. The market for reishi mushroom extract is segmented into organic and conventional reishi mushroom extracts. The organic extract is expected to draw significant attention among health conscious consumers in developed markets and subsequently act as a driving factor for reishi mushroom extract growth. Reishi extracts are categorized according to their forms, which include liquid extract and powdered extracts. The powdered extract segment is expected to build into a steady increase in market demand because of its diverse uses. Reishi mushroom extract is also segmented on the basis of end user applications, which include dietary supplements and cosmetics (Persistence Market Research 2018); in this regard, the end user industries for reishi mushrooms include the food and beverage, healthcare, and personal care industries. The geographical breakdown of the market consists of North America, Europe, Asia-Pacific, and Latin America, the Middle East and Africa (Allied Market Research 2018).

Medicinal products based on reishi mushroom extracts prevent tumor growth and fight cancer according to some scientific studies. Because cancer is one of the leading causes of mortality, there is great potential for using this type of extract in the global market. Furthermore, reishi mushrooms promote liver function, relieve stress, enhance immunity, and promote heart health and hormonal balance.

All these factors are expected to drive the market for reishi mushroom extract globally during the coming years (Persistence Market Research 2018). However, the main barriers to using reishi extract are its associated side effects, such as itchy throat, bloody stools, nosebleeds, liver toxicity, and diarrhea, as well as its low market penetration in the markets of developing countries.

The major market players in the market for reishi mushroom extracts are as follows (Allied Market Research 2018): Amax NutraSource, Inc., Bio-Botanica Inc., Bristol Botanicals Limited, Dragon Herbs, Hokkaido-reishi, Huachengbio, Mushroom Science, Nammex (North American Medicinal Mushroom Extracts), Qingdao Dacon Trading Co., Ltd., and Xian Yuensun Biological Technology Co., Ltd.

In the United States, the Food and Drug Administration (FDA) does not regulate the marketing of fungal medicinal products. As with other herbal supplements, the lack of regulation can create a "buyer beware" market because the integrity of the product is the responsibility of the manufacturer. With the current rapidly changing understanding of *Ganoderma* characteristics among species, even well-intentioned manufacturers may have difficulty ensuring that their product contains the labeled species (Loyd et al. 2018). Several Chinese species once called *G. lucidum* are now considered to belong to other species, including *Ganoderma flexipes, G. lingzhi, Ganoderma multipileum, Ganoderma sichuanense, G. sinense,* and *Ganoderma tropicum*.

Due to its various benefits in protecting human health, numerous *G. lucidum* products have emerged in the form of dietary supplements in the United States, which are also listed in the Dietary Supplements and Herbal Medicines of the United States Pharmacopeia (USP). However, Wu et al. (2017) showed that only 5 out of a sample 19 products labeled as *G. lucidum* dietary supplements purchased in the United States could be verified as genuine reishi mushrooms.

The authors employed a reliable and scientific toolkit, including a colorimetric assay, high performance thin layer chromatography (HPTLC), gas chromatography coupled with mass spectrometry (GC-MS), saccharide mapping based on polysaccharide analysis using carbohydrate gel electrophoresis (PACE), and high performance size exclusion chromatography coupled with a multi angle laser light scattering and refractive index detector (HPSEC-MALLS-RID) methods. These results suggest that the quality consistency of *G. lucidum* dietary supplements in the U.S. market was poor, and these inconsistencies should be carefully investigated. This situation can lead to a decrease in interest in medicinal mushrooms by the final consumers due to quality issues with the products on the market.

Market for Reishi Products in Colombia

In Colombia, the edible mushroom *Agaricus bisporus* (button mushroom) has been produced and marketed since the 1950s. Europeans who were exiled because of World War II started cultivating this edible mushroom in Latin American countries such as Colombia. The production of button mushrooms in Colombia increased from 800 tonnes per year in the 1990s to 8,000 tonnes per year today; however, there are different reports indicating that this country has consumed approximately 10,000 tons per year since 2010, according to Mushroom's Value (2014) and the company Fungus SAS of Santa Helena (Medellín, Colombia), a producer and marketer of these mushrooms.

Approximately 15 years ago, several species of edible mushrooms other than *A. bisporus* were produced in Colombia for commercial purposes. The main edible mushrooms cultivated for their medicinal attributes are *Pleurotus* spp., *L. edodes*, and *G. lucidum*. The cultivation of *Pleurotus ostreatus*, which is commonly called oyster mushroom in English, or orellana in Spanish, is the most popular in the different regions of Colombia, probably because it is a culture that can be maintained easily on a small scale, in an artisanal way. In regions such as Caldas, Cundinamarca, Antioquia, Córdoba, Cauca, and Huila, small oyster mushroom production facilities are in operation, in which 150 to 300 kg of these mushrooms are produced. According to the above information, between 3,000 and 6,000 kg of this fungus may be produced per month.

Orellana production is commercialized through sales to restaurants and, to some extent, to supermarkets. *L. edodes* production is frequently found in Cundinamarca, with an approximate production of 10,000 kg per month. This mushroom is on demand by gourmet restaurants around the country.

In Colombia, as in the rest of Latin America, the consumption of *G. lucidum* is recent compared to the consumption of *A. bisporus* mushrooms. The cultivation of the reishi mushroom has scarcely changed in 15 years in Colombia, a country that has experienced major developments related to the solid-state cultivation of edible and medicinal fungi other than button mushrooms. The demand for *G. lucidum* in Colombia has increased due to its medicinal attributes and the benefits of its constant consumption to human health. These benefits have been proven by different studies around the world, and Colombian consumers have become increasingly more aware of these characteristics. In addition, eating habits and the search for new products that offer an improvement in people's quality of life are booming, so products based on *G. lucidum* mushrooms are promising in these countries.

Small-scale commercial facilities for producing *G. lucidum* have been established in Colombia, such as the company Reishi de Colombia S.A.S., which is located in Chinchiná, Caldas. However, the production of this fungus is not sufficient to meet the current demand. In Medellín, the company Progal BT produces and markets products based on *G. lucidum* mycelium that are obtained by submerged fermentation.

In Colombia, edible products containing *G. lucidum* are sold, especially coffee, but the products are not available nationally. Table 2 lists the five companies with the highest production of coffee supplemented with *G. lucidum* in the world; the figures in the consumer satisfaction column correspond to the trust rankings published on the websites of these companies. The largest companies (both as producers and marketers) are DXN and Gano Excel from Malaysia. The marketing model of these companies corresponds to Internet or multilevel sales. These two companies offer different products based on coffee containing *G. lucidum*, but they have referenced few studies with scientific support.

Table 2. Comparison of the five largest companies in the world that produce coffee supplemented with *G. lucidum*

Company	Country	Existence (years)	Producer	Marketer	Consumer satisfaction (%)	Use of chemical additives
DXN	Malaysia	20	Yes	Yes	95	No
Gano Excel	Malaysia	18	Yes	Yes	100	Yes
Organogold	Canada	5	Alliance	Yes	14	Yes
Total Life Changes (TLC)	USA	2	No	Yes	N.R.	N.R.
Ganolife	Mexico	3	No	Yes	N.R.	N.R.

N.R.: Not reported.

Table 3. Specifications of the product "coffee with lactose-free milk and *G. lucidum*" in sachets (21 g)

Item	Organogold	Gano Excel	DXN
Coffee (%)	15.00	14.76	20.00
G. lucidum (%)	5.00	0.95	18.00
Nondairy milk (%)	37.00	41.67	30.00
Sugar (%)	43.00	42.62	32.00
Chemical additives	Yes	Yes	No
Grams per box	420	420	420
Dealer price (€)	16.50	16.50	9.50
Average price/10 g of *G. lucidum*	7.85	41.35	1.25
Qualitative valuation of *Ganoderma*	Expensive	Very expensive	Best price
100% organic	No	No	Yes

Source: Company websites.

One of the most frequently commercialized products is coffee with lactose-free milk and *G. lucidum* packed in sachets. The compositions of 21-g sachets of this product from three companies are listed in Table 3. It is clear that the product offered by the Gano Excel company contains 0.95% fungal extract, compared to the same product offered by the DXN company, which contains 18% *G. lucidum*; however, the label of the DXN company product does not specify if *G. lucidum* is from fungus powder or extract. The Organogold company states that its product contains 5% *G. lucidum* but also does not specify whether this percentage corresponds to a fungus powder or extract.

LIGNOCELLULOSIC MATERIALS AS FEEDSTOCKS

Lignocellulosic materials are classified according to their origin. Table 4 shows a general classification of the main materials with high lignocellulosic biomass contents. These materials are the important basic components needed for fungal development. Agricultural waste composed of straw, stover, husks, and other residues such as coffee pulp, and waste from vegetable, legume and oilseed crops, among others, can be used as raw materials to formulate substrates during the solid-state cultivation of *G. lucidum*. In particular, the most representative waste from coffee growing regions in tropical countries such as Colombia can be used for mushroom cultivation.

Table 4. Classification of the main materials with high contents of lignocellulosic biomass

Residues	Examples
Agricultural waste	Straw: Wheat, rice, barley, cotton, sorghum, millet, etc. Other waste: corn stover, coconut husks, waste cotton, coffee pulp, legume pods, sugarcane tops, coffee cut-stems
Agro-industrial waste	Bagasse: Sugarcane, sweet sorghum Husks: Rice, sorghum, corn, coffee, cocoa Oilseed cakes: Soybean, cotton seed, sunflower Other waste: Olive stone, silver film of coffee, coffee spent-ground, corn cobs, oil palm cake
Wood and forestry waste	Hardwood: Poplar, eucalyptus, oak, aspen, maple, birch, rosewood Softwood: Pine, fir, spruce, larch, cedar Sawdust, branches: oak, wood mixtures, residues from furniture production
Cellulosic residues	Newspaper, waste office paper, paper sludge
Herbaceous biomass	Grasses: Switchgrass, reed canary grass, coastal Bermuda grass, alfalfa hay, timothy grass
Municipal solid waste	Cellulosic fraction: Wastepaper, cardboard, fruit and vegetable peels, garden residues, wood items

Source: Modified from Cardona, Sánchez, and Gutiérrez (2010).

Herbaceous biomass comes from plants that do not generate wood, and their stems are green, such as the plants found in foliage, pastures, etc. Municipal solid waste is very heterogeneous and generally contains a high concentration of lignocellulosic materials, such as wastepaper, cardboard, fruit and vegetable peels, and garden waste, among others (Sánchez and Cardona 2007). The materials that are derived from wood are composed of two subgroups. The first subgroup corresponds to hardwood, which primarily includes species of angiosperms, such as poplar, eucalyptus, aspen, oak, and maple, among others. Softwood, in contrast, corresponds to coniferous wood (gymnosperm trees), such as pine, spruce, and larch, among others; softwood contains more lignin than hardwood. It should be highlighted that the largest proportion of timber forests are in the equatorial zone, that is, in the central and northern parts of South America, Central Africa and the Congo, the eastern coast of India, southern China and Indonesia. Thus, approximately 47% of these forests are found in the tropical zones of the planet (Henriksson, Brännvall, and Lennholm 2009).

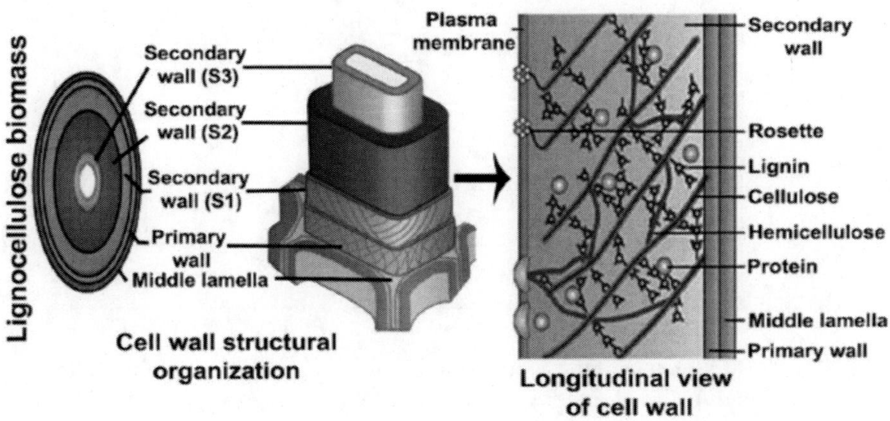

Figure 1. Diagrammatic structure and chemical composition of the lignocellulosic complex in plant cell walls. Source: Modified from Manavalan, Manavalan, and Heese (2015).

The lignocellulosic complex is common in plant cells since it is present in their cell walls, as shown in Figure 1. In fact, almost all the agricultural and agro-industrial residues employed for mushroom cultivation come from plant materials. In the figure, the arrangement of the cellulose microfibrils, along with the hemicellulose and lignin molecules that form the primary and secondary cell walls, is shown schematically. In general, lignocellulosic materials of natural origin are also known as lignocellulosic biomass. The main polymeric components of the lignocellulosic biomass are briefly described in the next subsections.

Cellulose

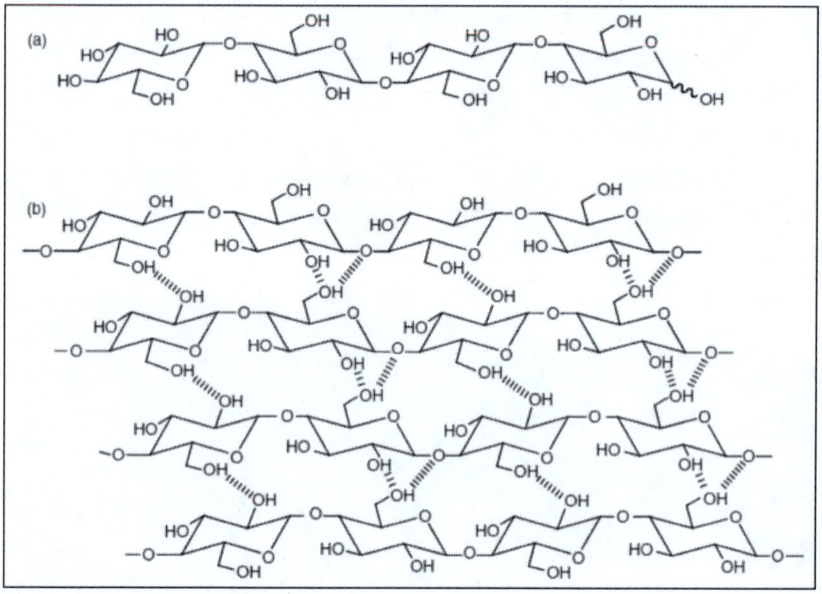

Figure 2. (a) Basic structure of the cellulose molecule. (b) Intramolecular illustration of the hydrogen bonds between cellulose molecules. Source: Izydorczyk, Cui, and Wang (2005).

Cellulose is a biopolymer composed of *D*-glucose units, and it is the most abundant organic molecule on earth. Plants synthesize cellulose as a structural material to support their weight. The longitudinal structures of cellulose, which are called microfibrils, form bundles via the hydrogen bonds that are created between the numerous OH groups of the glucose rings. In cellulose, *D*-glucose units are linked by β(1,4) glycosidic bonds, which give cellulose a very rigid and stable disposition. The basic structure of cellulose is shown in Figure 2.

The length of the cellulosic polymer is highly variable and depends on the organism from which the cellulose comes, as well as the age and metabolic status at the time of its extraction. An average degree of polymerization of approximately 10,000 was determined for wood cellulose (Zabel and Morrel 1992). Each glucose monomer presents a rotation of 180° with respect to the contiguous residues, and each one is stabilized by the formation of intramolecular hydrogen bonds. The microfibrils are formed by crystalline zones where the cellulose molecules are ordered and *para-*crystalline regions where they are disorganized (amorphous cellulose). The crystalline structure of cellulose makes the cell wall anisotropic (it has more than one index of light refraction as a function of the polarization direction of the light waves) and, therefore, has double refraction (birefringent) when viewed with polarized light (Meier 1955).

The crystalline (or native) cellulose is characterized by a high degree of ordering and polymerization, and thus it is insoluble, e.g., Avicel® (commercial microcrystalline cellulose), cotton fibers, filter paper, etc. Notably, modified cellulose is soluble and includes materials such as amorphous cellulose, carboxymethyl cellulose, and cello-oligosaccharides, in which the degrees of crystallinity and polymerization are lower (Levin 1998, Sánchez and Cardona 2007). Cellulose is a substance that is highly stable in response to chemical processes such as hydrolysis, in which high concentrations of acids or bases are required to break its structure successfully.

Microorganisms play an important role in the recycling of cellulosic carbon. There are a large number of microorganisms that are capable of breaking the cellulose units to monosaccharides under aerobic or anaerobic conditions, including the anaerobic bacteria *Bacteroides cellulosolvens*, *Bacillus* spp., *Clostridium celluloliticum*, *C. cellulovorans*, *C. papyrosolvens*, *C. thermocellum*, *Fibrobacter succinogenes*, *Cellvibrio gilvus*, and *Ruminococcus flavefaciens*; yeasts *Candida wickerhamii* and *C. lucitance*; ruminal fungi *Neocallimastix frontalis* (Joshi and Pandey 2009, Leschine 1995); and all the micro-filamentous aerobic fungi and macromycetes

(Carlile, Watkinson, and Gooday 2001). All these organisms use enzymes that are secreted to degrade the cellulose polymer, among which the endoglucanases, exocellobiohydrolases, and β-glucosidase (see below) should be highlighted, as well as the cellobiose kinase and phospho-β-glucosidase that may also be present in many of the organisms mentioned here.

Hemicellulose

Hemicelluloses are a heterogeneous group of polysaccharides that make up the cell wall of higher plants. These polysaccharides are generally linked by covalent and noncovalent bonds to cellulose and lignin. Hemicellulose is (in a generic way) amorphous, with little rigidity, and it is easily hydrolyzable in dilute acids or bases (Zabel and Morrel 1992). Hemicellulose makes up 20 to 40% of lignocellulosic biomass and consists of short, very branched chains of sugars (200 sugars on average). Among these sugars are, in order, *D*-xylose and *D*-arabinose (both 5-carbon sugars or pentoses), and *D*-galactose, *D*-glucose, and *D*-mannose (these latter sugars are 6-carbon sugars or hexoses). Other carbohydrate-related compounds, such as glucuronic, methyl glucuronic, and galacturonic acids, are also present in the hemicellulose structure. Furthermore, hemicelluloses contain, in a lower proportion, acetyl groups esterified to some of the OH groups of their different sugars (Cardona, Sánchez, and Gutiérrez 2010).

Hemicellulose involves different polysaccharides, the predominance of which determines their denomination. Thus, when *D*-xylose is the predominant monomer in a structure, a hemicellulose can be considered as a xylan. The structure of a hemicellulose can vary depending on its origin. The xylans are the main hemicelluloses in angiosperm wood (they constitute 10 to 35% of its dry weight) (Joselau and Ruel 1994, Zabel and Morrel 1992). Thus, for lignocellulosic materials derived from hardwood, the xylan backbone is composed of xylose units linked by β(1,4) bonds; from this backbone, very short branches come out, with usually only one sugar in length (Izydorczyk, Cui, and Wang 2005) through α(1,2) bonds with the methyl glucuronic acid (Figure 3a). In xylans from softwood, the acetyl groups are less frequent, but there are more branches due to the presence of α(1,3) bonds between the xylose backbone and arabino-furanose units (Cardona, Sánchez, and Gutiérrez 2010) (see Figure 3b). Considering its branched structure, hemicellulose does not form crystalline structures but, rather, forms amorphous structures. This biopolymer is more soluble in water and has higher susceptibility to hydrolysis (Hamelinck, Hooijdonk, and Faaij 2003).

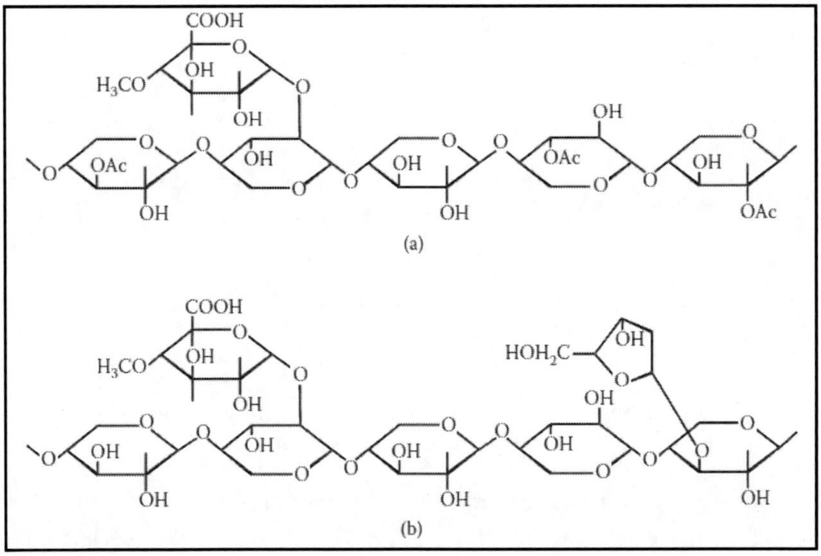

Figure 3. Partial structure of xylan chains corresponding to lignocellulosic materials obtained from hardwood (a) and softwood (b). Source: Cardona, Sánchez, and Gutiérrez (2010).

Hemicelluloses cover and unite cellulose microfibrils in a common matrix. The short extension of the chains that increases their solubility, and the exposed position of these chains on the surface of the microfibrils would explain why this polymer is among the first components of the cell wall to be attacked by the fungi that cause rot (Goodenough 1996, Zabel and Morrel 1992).

Lignin

Lignin is the most abundant and renewable aromatic material on the planet. Lignin comprises 10 to 25% of lignocellulosic biomass (Cardona, Sánchez, and Gutiérrez 2010), and it is an aromatic biopolymer that is not a polysaccharide. Lignin is insoluble in water and has a high molecular weight. This biopolymer is found primarily in the cell walls of plants, forming part of a matrix ordered along with the cellulose microfibrils.

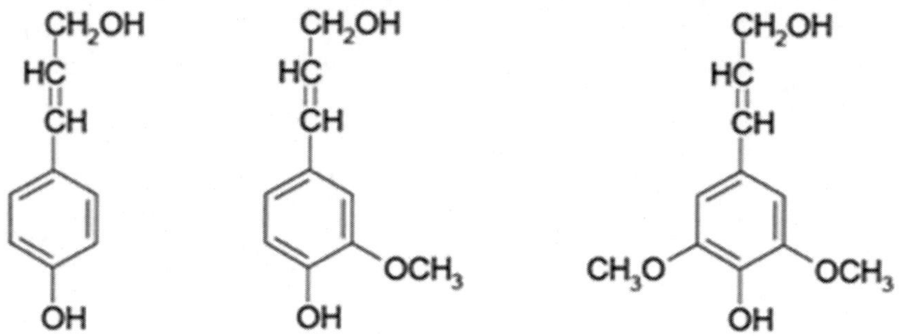

(a) (b) (c)

Figure 4. Cinnamyl alcohols (monolignols) are precursors of the lignin molecule. (a) Coumaryl alcohol. (b) Coniferyl alcohol. (c) Sinapyl alcohol.

Lignin is a very complex phenolic polymer made up of phenyl propane units linked by C-C and C-O-C bonds to form a three-dimensional amorphous structure (Lee 1997). The chemical precursors of lignin are the following cinnamyl alcohols: Sinapyl alcohol (3,5-dimethoxy-4-hydroxy cinnamyl alcohol), *p*-coumaryl alcohol (*p*-hydroxycinnamic alcohol), and coniferyl alcohol (4-hydroxy-3-methoxycinnamyl alcohol), which are shown in Figure 4. The structural units of lignin are based on these alcohols. Thus, the *p*-hydroxyphenyl units of lignin are derived from *p*-coumaryl alcohol, the guaiacyl units are derived from coniferyl alcohol, and the syringyl units are derived from sinapyl alcohol (Cardona, Sánchez, and Gutiérrez 2010, Zakzesky et al. 2010).

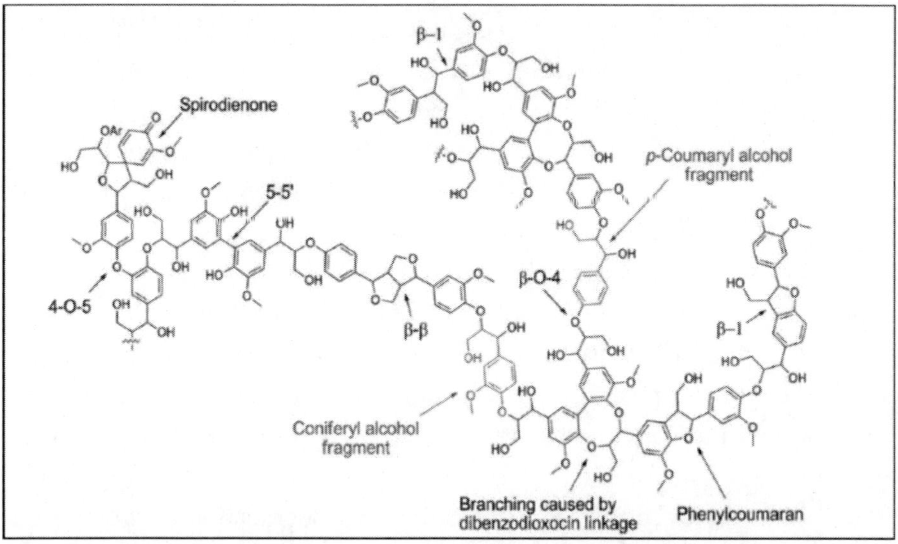

Figure 5. Schematic representation of the structure of lignin from softwood. Source: Zakzesky et al. (2010).

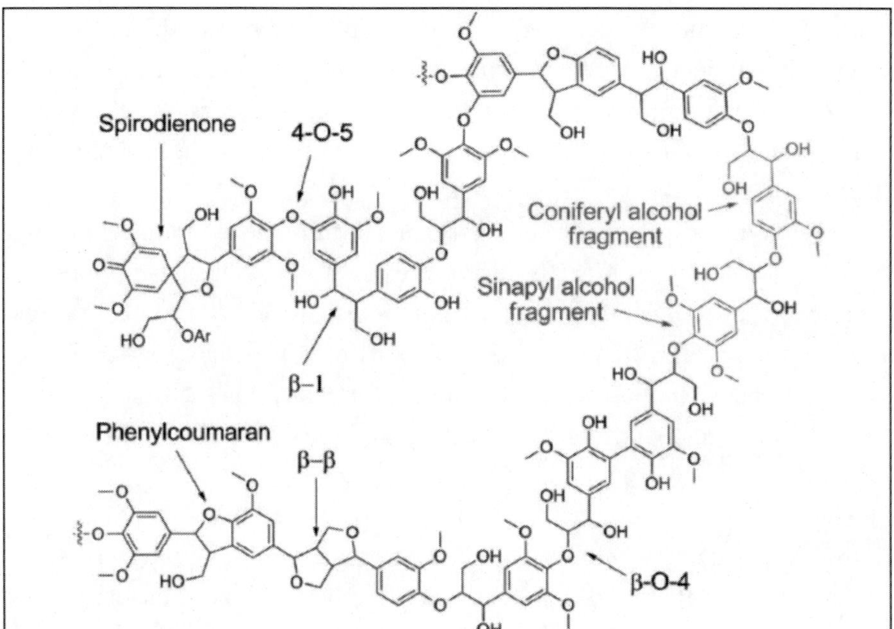

Figure 6. Schematic representation of the structure of lignin from hardwood. Source: Zakzesky et al. (2010).

The composition, molecular weight and amount of lignin differs from plant to plant with the abundance of lignin, generally decreasing in the following order: softwoods > hardwoods > grasses. Schematic representations of the lignin from softwoods and hardwoods are shown in Figure 5 and Figure 6, respectively. The composition of the lignin in softwood and hardwood is relative to the abundance of its three precursor alcohols. Softwood lignin is composed of 90% coniferyl alcohol, while an approximately equal proportion of coniferyl and sinapyl alcohol appear in hardwood lignin. Likewise, the methoxy groups in the aromatic rings prevent the formation of dibenzodioxocin linkages, which implies a more linear form in the structure of the hardwood lignin polymer than that of softwood.

Lignin provides resistance to the microbial degradation of wood. The intimate spatial association between polysaccharides and lignin within the cell walls of wood provides a protective barrier that prevents the degradation of cellulose. Once the lignin surrounding the cellulose fibrils is removed or modified, the cellulose becomes more accessible to the microbial enzymes and can be efficiently degraded. Thus, the lignification in many plants can be a mechanism of resistance to disease, and it occurs as a defense mechanism against pathogenic fungi or in response to wounds (Blanchette 1995).

To date, the only organisms that are able to mineralize lignin efficiently into CO_2 and water as final products are the fungi that cause white rot (Leonowicz et al. 1999). These fungi mostly belong to the orders Aphyllophorales and Agaricales (Peláez, Martínez, and Martínez 1995) and have a complex of oxidases and peroxidases for catalyzing the first reactions that break linkages within the complex lignin molecule, generating smaller molecules.

The degradation products are then incorporated into the metabolic cycles of the organism with CO_2 as the end product (Kirk and Farrel 1987). It has been shown that using lignin as a single substrate is not a usable source for the growth of fungi, but that the presence of a cosubstrate is necessary (Kirk, Connors, and Zeikus 1976).

Other Polymers Contained in Lignocellulosic Materials

Starch and pectins may be present in lignocellulosic materials, depending on the source. Starch is a homopolysaccharide of *D*-glucose units bound by α(1,4) glycosidic bonds. Starch has two components, amylose and amylopectin. Amylose is a linear molecule with a molecular weight close to 5,000 Da and is naturally composed of *D*-glucose linked by α(1,4) bonds. Amylopectin is a polysaccharide with α(1-4) and α(1-6) bonds every 25-30 glucose units. Its molecular weight is different for different sources of starch.

Starch is widely distributed in microalgae, algae, and higher plants, and it is the best storage material. For heterotrophic organisms, starch is the most common source of energy.

A large number of microorganisms have the ability to use this polysaccharide as an energy and carbon source (Joshi and Pandey 2009). In biological systems, the hydrolysis of starch into glucose is performed by multiple enzymes, such as α-amylase, β-amylase, glucoamylase, and maltose hydrolase (β-amylase), as well as other enzymes, such as amylo-1-6-glucosidase, limit-dextrinase, and pullulanases (Cardona, Sánchez, and Gutiérrez 2010, Hyun and Zeikus 1985, Takasaki 1976).

Pectins are another important component of the plant cell wall that have a polysaccharidic nature. They are one of the main constituents of soluble fiber. Pectins consist mainly of *D*-galacturonic acid, a derivative of *D*-galactose in which the hydroxyl group of carbon 6 has been oxidized to form a carboxylic group. These materials are also composed of *D*-rhamnose, *D*-arabinose, and *D*-galactose, although to a lesser extent. The commercial importance of pectins is based on their use as a gelling agent during the processing of various foods, such as yogurts, preserves, jellies, and jams (Campbell and Farrell 2004). Pectins play an essential role in the stability of the cell wall by acting as a binding material for cellulose fibers (Pasquel 2001).

Formulation of Substrates Based on Lignocellulosic Materials

The dry base composition of different lignocellulosic materials is presented in Table 5. As shown, cellulose is present in all the materials indicated in this table. In particular, the cellulose and hemicellulose contents of hard and soft woods, sugarcane bagasse, paper waste, cereal straws (in this case, wheat), corn cobs and grasses are very significant. These high cellulose contents make these materials relevant to obtaining mushrooms and other value-added products by fermenting the sugars derived from the hydrolysis of this polysaccharide. The lignin contents are variable in the materials shown in Table 5, but they can be treated with white rot fungi, which are producers of lignin-degrading enzymes (oxido-reductases) as discussed by Kirk and Farrel (1987).

Table 5. Percentage composition (dry basis) of some lignocellulosic materials

Materials	Cellulose (% by weight)	Hemicellulose (% by weight)	Lignin (% by weight)
Sugarcane bagasse	50	25	25
Hardwood	40-55	24-40	18-25
Softwood	45-50	25-35	25-35
Nut shells	25-30	25-30	30-40
Corn cobs	45	35	15
Grasses	25-40	35-50	10-30
Paper	85-99	0	0-15
Wheat straw	30	50	15
Sorted refuse	60	20	20
Leaves	15-20	80-85	0
Cotton seed hairs	80-95	5-20	0
Newspaper	40-55	25-40	18-30
Wastepaper from chemical pulp	60-70	10-20	5-10
Primary wastewater solids	8-15	0	24-29
Coastal Bermuda grass	25	36	6
Switchgrass	45	31	12
Empty fruit bunches, oil palm	15	12	7
Palm press fiber	24	14	13

Sources: Cardona, Sánchez, and Gutiérrez (2010), Sun and Cheng (2002).

Table 6. Main organic materials derived from agro-industrial chains that process important crops in Colombia according to their regional distribution

Crop	Materials	Department	Ligno-cellulosic	Proteic
Sugarcane	Sugarcane bagasse from sugar mills and noncentrifugal sugar facilities, molasses, sugarcane tops and leaves	Valle, Caldas, Santander, Antioquia	X	
Coffee	Used coffee grounds, husks, pulp, mucilage, cut-stems	Caldas, Risaralda, Quindío, Antioquia, Tolima, Nariño, Huila	X	
Cereals	Husks, straws and other residues	Tolima, Antioquia, Valle, departments at Eastern plains	X	
Legumes	Waste from cropping and harvesting	Tolima, Valle, Caldas, Quindío, Risaralda, Antioquia Santander, Cundinamarca	X	X
Oilseeds	Harvesting waste and cakes	Tolima, Antioquia, departments at Atlantic coast, Valle	X	X
Woods	Sawdust, chips	Caldas, Antioquia, Cundinamarca, Chocó	X	
Fruit trees	Peels, seeds, harvesting waste	The whole country	X	
Vegetables	Harvesting waste, leaves, stems, peels	Caldas, Cundinamarca, Antioquia	X	
Cocoa	Husks, cropping waste	Caldas, Antioquia, Cundinamarca	X	X

The raw materials used more frequently during cultivation processes are agricultural and agro-industrial residues due to their high availability and low price. In Colombia, a tropical country with high agricultural activity, there is a large amount of waste that can be used as feedstock for substrate formulation. Table 6 shows the regional distribution of the main lignocellulosic materials derived from different economically important crops and related agro-industrial chains in Colombia; some proteic materials are also presented. The administrative regions in Colombia are called departments (the equivalents of provinces). Figure 7 shows the primary feedstocks available in the department of Caldas that were employed in substrate preparation for mushroom production.

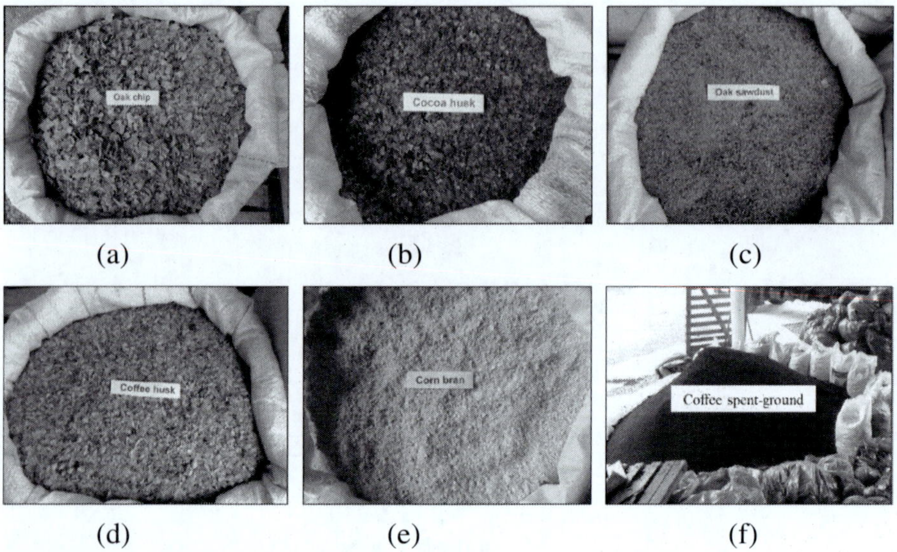

Figure 7. Some representative feedstocks for substrate preparation available in the Caldas region. (a) Oak chips. (b) Cocoa husks. (c). Oak sawdust. (d) Coffee husks. (e). Corn bran. (d). Coffee spent-ground.

In the following research case, the selection of feedstocks to formulate culture medium for cultivating *G. lucidum* mushrooms is briefly described.

Research Case 1. Feedstock Selection and Development of Formulations for Cultivating G. lucidum Mushrooms

In the Bioprocess and Agro-industry Plant at the Universidad de Caldas, the raw materials used in Colombia for cultivating white rot fungi, such as *G. lucidum*, have been selected. These materials are mostly agricultural and agro-industrial wastes. Table 6 indicates the main lignocellulosic residues available in the different regions of Colombia used for the cultivation of these fungi. The materials used as feedstocks for substrate formulation to cultivate *G. lucidum* were selected according to their availability by region and by accounting for the nutritional requirements of the fungus. The carbon/nitrogen (C/N) ratio of the mixture of materials used as a culture medium is a key parameter to consider since it directly influences the growth and development of the fungi. For the calculation of this ratio, the total nitrogen was determined using the Kjeldahl (1883) method; for the carbon content, the ash contents in the samples were determined by using the method described in Colombian Technical Standard NTC5167 (ICONTEC 2011), which was based on the AOAC analytic procedures to quantify the volatiles content; and the carbon content was assumed to be 55-58% of the volatile contents present in the samples. In this case, the C/N ratios that generated appropriate results for the growth and development of the *G. lucidum* fungus were between 40 and 80. For this culture, multiple tests were performed with different lignocellulosic residues and different proportions on a dry basis. These multiple mixtures were monitored during the growth and development of the mushrooms during both the vegetative and reproductive phases. The criterion employed to assess the performance of the cultivation process was the biological efficiency (*BE*), which indicates the mass of the dry substrate at the time of inoculation by using the following equation:

$$BE = \frac{mass\ of\ fresh\ mushrooms}{mass\ of\ dry\ substrate\ at\ the\ time\ of\ inoculation} \times 100\% \quad (1)$$

The cultivation tests for reishi production began in 2008 at the former Macromycetes Laboratory at the Universidad de Caldas (in 2012, the equipment and materials from this laboratory were moved to the new facilities at the Bioprocess and Agro-industry Plant) using a strain donated by the Laboratory of Experimental Mycology at the Universidad de Buenos Aires (Argentina). The first formulation (on a dry basis) that was tested was as follows: 80% oak sawdust, 17% corn bran, 2% $CaCO_3$, and 1% $CaSO_4$, with an initial substrate moisture of 60%. This formulation was tested on a small scale by using 20 kg of substrate packed into 1-kg plastic bags. The resulting *BE* was 8%. The mushrooms were of good quality, but the *BE* was too low to be competitive for future commercial production.

Table 7. Selected formulations of substrates for reishi cultures exhibiting high biological efficiencies

Composition (% dry basis)	Sawdust	Corn bran	Coffee husks	Gypsum ($CaSO_4$)	$CaCO_3$	Cane sugar
Formulation 1	77	20	-	1	1	1
Formulation 2	50	17	29	-	2	2
Formulation 3	50	23	24	-	2	1

Subsequently, new *G. lucidum* strains were acquired and a series of formulations were proposed. These formulations were first tested on a small scale. The mixtures with the best performance were selected for testing at the pilot scale (in cultivation rooms with a capacity of 200 kg of substrate). These formulations are shown in Table 14 (see Chapter 2). With the proposed formulations, increases in the *BE* of up to 30% were reached at the pilot scale. These results allowed for the development of prototypes for *G. lucidum* cultivation at a pilot scale for industrialization purposes since these *BE* values are competitive for commercial purposes. In addition, the quality of the *G. lucidum* fruiting bodies obtained from these formulations was standardized. Some selected and refined formulations of the ones that have been tested in the Bioprocess and Agro-industry Plant at the Universidad de Caldas are presented in Table 7.

Final Considerations

Ganoderma lucidum is a mushroom that is currently in high demand around the world due to its medicinal attributes. Its consumption is projected to increase significantly around the world in the coming years. In tropical countries such as Colombia, its demand and consumption as a minimally processed mushroom is increasing, especially when it is dried and ground and mixed with food. For mushroom production, there are many residual materials from different agricultural and agro-industries activities that can be employed. These materials are primarily lignocellulosic waste, which implies an additional benefit for the environment since their utilization represents a way to take economic advantage of this waste to produce value-added products.

REFERENCES

Allied Market Research. 2018. "*Reishi Mushroom Extract Market - Global Opportunity Analysis and Industry Forecast, 2017-2023.*" Allied Market Research. Accessed December 2018. https://www.alliedmarketresearch.com/reishi-mushroom-extract-market.

Blanchette, R. A. 1995. "Degradation of the lignocellulose complex in wood." *Canadian Journal of Botany* 73:S999-S1010.

Campbell, M. K., and S. O. Farrell. 2004. *Bioquímica*. México: International Thomson Editores. [*Biochemistry*. México: International Thomson Publishers].

Cao, Y., S.-H. Wu, and Y.-C. Dai. 2012. "Species clarification of the prize medicinal *Ganoderma* mushroom "Lingzhi"." *Fungal Diversity* 56 (1):49-62.

Cardona, C. A., Ó. J. Sánchez, and L. F. Gutiérrez. 2010. *Process Synthesis for Fuel Ethanol Production*. Boca Raton, FL, USA: CRC Press.

Carlile, M. J., S. C. Watkinson, and G. W. Gooday. 2001. *The Fungi*. London: Academic Press.

Chang, S. T., and J. A. Buswell. 1996. "Mushroom nutriceuticals." *World Journal of Microbiology and Biotechnology* 12:473–76.

Chang, S. T., and P. G. Miles. 2004. *Mushrooms. Cultivation, Nutritional Value, Medicinal Effect, and Enviromental Impact*. London: CRC Press.

Chang, S. T., and S. P. Wasser. 2012. "The role of culinary-medicinal mushrooms on human welfare with a pyramid model for human health." *International Journal of Medicinal Mushrooms* 14 (2):93-134.

Chang, S. T., and S. P. Wasser. 2017. The Cultivation and Environmental Impact of Mushrooms. In *Oxford Research Encyclopedia of Environmental Science*. Oxford: Oxford University Press USA.

Chen, R. Y., and D. Q. Yu. 1991. "Studies on the triterpenoid constituents of the spores from *Ganoderma lucidum*." *Acta Pharmaceutica Sinica B* 26:267-73.

CNBC. 2018. "*Global Mushroom Market expected to reach USD 59.48 billion in 2021*." CNBC LLC. Accessed December 2018. https://www.cnbc.com/2016/10/26/globe-newswire-global-mushroom-market-expected-to-reach-usd-5948-billion-in-2021.html.

FAO. 2018. "*FAOSTAT*." Food and Agriculture Organization of the United Nations (FAO). Accessed December 2018. http://www.fao.org/faostat/en/#data/QC.

Goodenough, P. W. 1996. "Structural studies on cellulases, pectinases an xylanases." In *Enzymes for Carbohydrate Engineering*, edited by K.-H. Park, J. F. Robyt and Y.-D. Choi, 83-107. Amsterdam: Elsevier.

Hamelinck, C. N., G. v. Hooijdonk, and A. P. C. Faaij. 2003. "*Prospects for ethanol from lignocellulosic biomass: techno-economic performance as development progresses*." Report NWS-E-2003-55. Utrecht, the Netherlands: Utrecht University.

Henriksson, G., E. Brännvall, and H. Lennholm. 2009. "The Trees." In *Wood Chemistry and Biotechnology*, edited by M. Ek, G. Gellersted and G. Henriksson, 13-44. Berlin: De Gruyter.

Hyun, H. H., and J. G. Zeikus. 1985. "General biochemical characteriza-tion of thermostable extracellular β-amylase from *Clostridium thermosulfurogenes*." *Applied and Environmental Microbiology* 49 (5):1162-67.

ICONTEC. 2011. NTC 5167. Productos para la industria agrícola. Productos orgánicos usados como abonos o fertilizantes y enmienda o acondicionadores de suelo. In *NTC 5167*. Bogotá: Instituto Colombiano de Normas Técnicas y Certificación (ICONTEC).

Izydorczyk, M., S. W. Cui, and Q. Wang. 2005. "Polysaccharide Gums: Structure, Functional Properties, and Applications." In *Food Carbohydrates*, edited by S. W. Cui, 293-99. Boca Raton, FL, USA: CRC Press, Taylor & Francis.

Joselau, J. P., and K. Ruel. 1994. "Wood polysaccharides and their degradation by fungi." In *Host Wall Alterations by Parasitic Fungi*, edited by O. Petrini and G. B. Ouellette, 334-87. St. Paul, MN, USA: APS Press.

Joshi, V. K., and A. Pandey. 2009. *Biotechnology: Food Fermentation. Microbiology, Biochemistry and Technology*. Vol. 1. New Delhi: Educational Publishers & Distributors.

Kim, H. W., and B. K. Kim. 2002. "Recent advances on the biologycally active triterpenoids of *Ganoderma lucidum*." In *Ganoderma: Genetics, Chemistry, Pharmacology and Therapeutics*, edited by Z.-B. Lin, 10-19. Beijing: Medical University Press.

Kirk, T. K., W. J. Connors, and J. G. Zeikus. 1976. "Requeriment for a growth substrate during lignin decomposition by two wood-rotting fungi." *Applied and Environmental Microbiology* 32:192-94.

Kirk, T. K., and R. L. Farrel. 1987. "Enzymatic combustion: the microbial degradation of lignin." *Annual Reviews of Microbiology* 41:465-505.

Kjeldahl, J. 1883. "Neue methode zur bestimmung des stickstoffs in organischen körpern (New method for the determination of nitrogen in organic bodies, in German)." *Zeitschrift für analytische Chemie* 22 (1):366-82.

Lee, J. 1997. "Biological conversion of lignocellulosic biomass to ethanol." *Journal of Biotechnology* 56:1-24.

Leonowicz, A., A. Matuszewska, J. Luterek, D. Ziegenhagen, M. Wojtaś-Wasilewska, N.-S. Cho, M. Hofrichter, and J. Rogalski. 1999. "Biodegradation of lignin by white-rot fungi." *Fungal Genetics and Biology* 27 (2-3):175-85.

Leschine, S. B. 1995. "Cellulose degradation in anaerobic environments." *Annual Reviews in Microbiology* 49:399-426.

Levin, L. 1998. "Biodegradación de materiales lignocelulósicos por *Trametes trogii* (*Aphyllophorales*, *Basidiomycetes*)." PhD Diss., Facultad de Ciencias Exactas y Naturales, Universidad de Buenos Aires. ["Biodegradation of lignocellulosic materials by *Trametes trogii* (*Aphyllophorales*, *Basidiomycetes*)." PhD Diss., Faculty of Exact and Natural Sciences, Universidad de Buenos Aires].

Loyd, A. L., B. S. Richter, M. A. Jusino, C. Truong, M. E. Smith, R. A. Blanchette, and J. A. Smith. 2018. Identifying the "Mushroom of

Immortality": Assessing the Ganoderma Species Composition in Commercial Reishi Products. *Frontiers in Microbiology* 9 (1557). Accessed December 2018. doi:10.3389/fmicb.2018.01557.

Manavalan, T., A. Manavalan, and K. Heese. 2015. "Characterization of lignocellulolytic enzymes from white-rot fungi." *Current Microbiology* 70 (4):485-98.

Meier, H. 1955. "Decomposition of the cell wall by wood-destroying fungi and the submicroscopic structure of tracheids of spruce and wood fibers of birch." *Holz als Rohund Werkstoff* 13:323-38.

Mushroom's Value. 2014. "*Información Económica.*" Mushroom's Value S.L. Accessed December 2018. http://www.mushroomsvalue.com/informacion-economica/.

Papinutti, V. L., L. A. Dorio, and F. Forchiassin. 2003. "Degradación de madera de álamo por *Fomes sclerodermeus*: Producción de enzimas ligninolíticas en aserrín de álamo y cedro." *Revista Iberoamericana de Micología* 20:16-20. ["Degradation of poplar wood by *Fomes sclerodermeus*: Production of ligninolytic enzymes in poplar and cedar sawdust." *Iberoamerican Journal of Mycology* 20:16-20].

Pasquel, A. 2001. "Gomas: una aproximación a la industria de alimentos." *Revista Amazónica de Investigación Alimentaria* 1 (1):1-8. ["Gums: An approach to the food industry." *Amazon Journal of Food Research* 1(1):1-8].

Peláez, F., M. J. Martínez, and A. T. Martínez. 1995. "Screening of 68 species of *basidiomycetes* for enzymes involved in lignin degradation." *Mycological Research* 99 (1):37-42.

Persistence Market Research. 2018. "Reishi Mushroom Extract Market: Global Industry Trend Analysis 2012 to 2017 and Forecast 2017 - 2025." *Persistence Market Research.* Accessed December 2018. https://www.persistencemarketresearch.com/market-research/reishi-mushroom-extract-market.asp.

Rajarathnam, S., and M. N. Shashirekha. 2003. "Mushrooms and Truffles. Use of Wild Mushrooms." In *Encyclopedia of Food Sciences and Nutrition (Second Edition)*, edited by B. Caballero, 4048-54. Oxford: Academic Press.

Romm, A., I. Burgess, D. Winston, S. M. Zick, and A. M. Crawford. 2010. "Conditions of the Reproductive Organs." In *Botanical Medicine for Women's Health*, edited by A. Romm, M. L. Hardy and S. Mills, 211-55. Saint Louis: Churchill Livingstone.

Sánchez, Ó. J., and C. A. Cardona. 2007. *Producción de Alcohol Carburante: Una Alternativa para el Desarrollo Agroindustrial*. Manizales, Colombia: Universidad Nacional de Colombia. [*Fuel Ethanol Production: An Alternative for Agro-industrial Development*. Manizales, Colombia: Universidad Nacional de Colombia].

Stamets, P. 2000. *Growing Gourmet and Medicinal Mushrooms*. Berkeley: Ten Speed Press.

Sun, Y., and J. Cheng. 2002. "Hydrolysis of lignocellulosic materials for ethanol production: A review." *Bioresource Technology* 83:1-11.

Takasaki, Y. 1976. "Production and utilization of β-amylase and pullulanase from *Bacillus cereus* var. *mycoides*." *Agricultural and Biological Chemistry* 40 (8):1523-30.

Wasser, S. P. 2010. "Medicinal mushroom science: History, current status, future trends, and unsolved problems." *International Journal of Medicinal Mushrooms* 12 (1):1-16.

Wu, D.-T., Y. Deng, L.-X. Chen, J. Zhao, A. Bzhelyansky, and S.-P. Li. 2017. "Evaluation on quality consistency of *Ganoderma lucidum* dietary supplements collected in the United States." *Scientific Reports* 7 (1):7792.

Zabel, R. A., and J. J. Morrel. 1992. *Wood Microbiology: Decay and Its Prevention*. San Diego: Academic Press.

Zakzesky, J., P. C. A. Bruijnincx, A. L. Jongerius, and B. M. Weckhuysen. 2010. "The catalytic valorization of lignin for the production of renewable chemicals." *Chemical Reviews* 110 (6):3552-99.

Zion Market Research. 2018. "*Mushroom Market to Witness Fabulous Growth in the With A CAGR of 9.2%.*" Zion Market Research. Accessed December 2018. https://www.zionmarketresearch.com/news/global-mushroom-market.

Chapter 2

SOLID-STATE CULTIVATION OF *GANODERMA LUCIDUM*

INTRODUCTION

The cultivation techniques for *G. lucidum* mushrooms have evolved considerably over the last 50 years. Chinese researchers have significantly contributed to these advances. These investigators have developed different solid-state artificial techniques and submerged fermentation methods for producing mycelial biomass and metabolites of pharmacological interest. In the West, this culture system has been subject to demand at greater scales, especially during the last decade. The production of this species at larger scales allows for the use of its fruiting bodies as a final product (fresh mushrooms) and as a feedstock for the production of nutraceutical products, nutritional supplements, and others. For this purpose, procedures for extracting and refining bioactive substances from the mycelium and fruiting bodies of *G. lucidum* have been developed in addition to diverse products containing this mushroom or its parts, such as enriched nutritional supplement beverages, injectable fluids, pills, and others (Stamets 1993).

G. lucidum grows naturally on a wide variety of substrates and mainly on dead trees, such as oak, elm, maple, willow and plum trees, among others, making it one of the fungi that cause white rot in wood (Stamets 1993). For the commercial production of the fungus, it was necessary to develop formulations of enriched substrates. From this effort, the artificial solid culture of *G. lucidum* has emerged. For this culture, different lignocellulosic materials are used, which in general are diverse agricultural and agroindustrial residues. An appropriate formulation of the solid media (substrates) allows for the enhancement of the process yield that is related to an increase in biological efficiency and a decrease in the process time. In addition, the implementation of these artificial cultures for various macromycete species requires the user to have control over the process, from the pretreatment of feedstocks (when necessary) to the controlled environment of the cultivation rooms.

The solid-state cultivation of macromycetes has advantages, such as the utilization of different types of agricultural and agroindustrial waste that cause pollution when they are underused or not used and the establishment of the intensive production of mushrooms with high nutritional and medicinal value that, in turn, can enhance the human immune system and improve the quality of life of people. Moreover, this type of cultivation technique generates positive socioeconomic impacts by improving the economic situation of the involved communities (Chang 2009).

The materials used for the solid-state cultivation of *G. lucidum* in China before 1983 were based on wood residues and tree trunks. However, the development of substrates based on giant grasses supplemented with minerals and cereal derivatives for the cultivation not only of this species but also of other macromycetes began in 1987 (Lin and Lin 2014); that development led to similar biological efficiency results and expanded the alternatives to cultivation in poor environments, such as wildlands, desert areas or areas where cultivation was not feasible, by utilizing a wide range of agricultural and agroindustrial wastes to formulate the solid substrates required to cultivate this mushroom.

BIOLOGY AND FEATURES OF *GANODERMA LUCIDUM*

Biology and Taxonomy of *G. lucidum*

G. lucidum (W. Curt.: Fr.) P. Karst belongs to the phylum Basidiomycota, class Basidiomycetes, order Aphyllophorales, family Polyporaceae (or Ganodermataceae), and genus *Ganoderma*. Although there are many *Ganoderma* spp., the most often cultivated species in this genus at the commercial level are *Ganoderma lucidum*, *G. capense*, and *G. sinense* (Chang and Miles 2004). The taxonomy of *Ganoderma* species has usually been based on classic description criteria, which are not currently accepted in the global scientific literature. In recent decades, several alternative methods for the taxonomic study of *Ganoderma* spp. have become available, such as sequencing analysis, ribosomal genes and spacers, and rDNA analysis (Gottlieb, Ferrer, and Wrigth 2000). However, more analyses of this type are needed to complete the taxonomic description of the *Ganoderma* genus and the detailed description of the *G. lucidum* species.

G. lucidum is a nonedible fungus because it tastes bitter and has a woody consistency, but it has been widely used in traditional Eastern medicine since ancient times as a source of longevity and as a health promoter. Its fruiting body is known as *lingzhi* (celestial herb) in China, *youngzhi* in Korea, *reishi* in Japan, and only "ganoderma" in the USA (Chen and Miles 1996, Leskosek-Cukalovic et al. 2010).

Morphological Features of *G. lucidum*

The mycelium of *G. lucidum* is light and white, with hyphae measuring 1-3 μm in diameter. There are morphological and structural differences in the different parts of the mycelium. The linear cells at the tip of the hyphae are narrow, at only 1 μm in diameter. The cell walls are thin and transparent, and the protoplasm is uniform and dense. The active part that generates movement in the mycelium and its elongation corresponds to the tip of the hypha. The middle cells of the hyphae are slightly thicker, with thicker cell walls and a sparser cytoplasm. The linear end cells are comparatively thicker and contain dispersed granular material (Lin and Lin 2014).

The fruiting bodies of *G. lucidum* measure 5-20 cm and have a shiny surface that gives the impression of being varnished when they are fresh. The carpophores are formed by a tissular or organized mycelium, which make up the stipe, pileus, and hymenium. The stipe or stem can be lateral or central and irregularly cylindrical. The pileus or cap is reniform and semicircular, but rarely circular. The fruiting bodies are fleshy, corky and faintly yellowish during their formation, and they become brownish during their growth until they reach a reddish or dark brown color with different tones spanning up to the margin when they reach maturity. The margin is usually thin and slightly curved downwards. The pores of *G. lucidum* fruiting bodies are white in the young bodies, turning yellowish with the maturity of the fungus and sporulation. The spores are brown, oval, and 8.5-11.5 × 5.0-7.0 μm in size, and they are contained in a soft inner sac with a rugged external appearance (Ying et al. 1987).

Generalities of the Nutritional Requirements for Fungal Growth and Development

Fungal growth is a complex phenomenon. This growth can be defined as the ordered increase in the cellular components involved in increases in the biomass. The rates of cell growth can vary due to different factors, such as the genetic characteristics and the number of generations (multiplications) of the strain. In addition, external factors also influence the growth of the fungi, such as the environmental conditions, the physical-chemical characteristics and state (solid or liquid) of the substrates in which they grow, and the aseptic conditions of the production facilities where the mushrooms are produced, among others. These factors also directly or indirectly affect the production of valuable metabolites during the different phases of fungal growth.

Additionally, during the reproductive phase of the fungi, multiple specialized structures are involved, and there is a high demand for nutrients that must be present in the substrates, especially when the mushroom cultivation is expected to be efficient during the mass production of fruiting bodies.

Among the nutritional requirements of fungi, it is necessary to consider both the oxygen requirements and the hydrogen ion concentration (pH). Although there are fungi, such as yeasts, that can live under anaerobic conditions, most fungi are considered aerobic. Low concentrations of oxygen in the media where the fungi develop can have a negative effect due to the oxygen deficit. In addition, the pH can cause positive and negative effects on the morphological development of the fungus, as this factor is a determinant for the absorption and assimilation of nutrients. In general, most fungi develop properly at pH ranges between 4 and 8, although there are exceptions, and therefore, these species should be handled separately. The oxygen supply and pH directly influence the metabolic processes and consequently the ability of the fungal species to use the nutrients from the medium within a short time (Carlile, Watkinson, and Gooday 2001, Chang and Miles 2004, Hudson 1986).

Carbon provides the energetic and structural requirements of fungal cells. Carbon is taken up by fungi from various sources, such as polysaccharides (such as cellulose and hemicellulose), monosaccharides, organic acids, amino acids, alcohols, and polycyclic compounds; in addition, they have the ability to degrade other polymeric substances such as lignin. Nitrogen is essential for the synthesis of proteins, purines, and pyrimidines. Chitin, a polysaccharide that is common in the cell walls of many fungi, also contains nitrogen. Fungi use a variety of sources to obtain the necessary nitrogen for synthesizing these essential compounds (Carlile, Watkinson, and Gooday 2001, Montoya 2012).

Minerals and vitamins are also very important for the growth and development of fungi, in addition to their direct role during the production of primary and secondary fungal metabolites.

Table 8. Required vitamins and minerals for fungal development

Nutrient	Supply form	Function	Average content
Minerals			
Phosphorus (P)	Potassium phosphate or phosphorus-containing materials	Macroelement; synthesis of ATP, nucleic acids, and phospholipids of cell membranes	10^{-4} M
Potassium (K)	Potassium phosphate	Macroelement; cofactor of some enzymatic systems; involved in the metabolism of carbohydrates; important for maintaining the ionic balance of fungi	10^{-3} M
Magnesium (Mg)	Magnesium sulfate	Macroelement; many enzymes are activated by magnesium; important for ATP metabolism	10^{-3} M
Iron (Fe)	Added salts or present in substrates	Microelement; activator of oxidoreductase enzymes	10^{-3} M
Zinc (Zn)	Added salts or present in substrates	Microelement; activator of enzymes such as alcohol dehydrogenase	10^{-6} M
Manganese (Mn)	Added salts or present in substrates	Microelement; activator of enzymes such as oxidoreductases; synthesis of ATP and nucleic acids	10^{-8} M
Copper (Cu)	Added salts or present in substrates	Microelement; activator of many enzymes; toxic for fungi in high concentrations	10^{-6}–10^{-7} M
Molybdenum (Mo)	Added salts or present in substrates	Microelement; required if the nitrogen source contains nitrates, as this element is a constituent of the flavoprotein nitrate reductase, which reduces nitrates into ammonium ions	10^{-7} M
Calcium (Ca)	Added salts or present in substrates	Microelement; required for carpophore formation in several macromycetes species	10^{-6} M (very variable)
Vitamins			
Thiamine (vit. B_1)	Added as pyrimidines or thiazoles; some species require the direct addition of thiamine	Coenzyme of carboxylase for converting pyruvic acid into acetaldehyde and carbon dioxide during carbohydrate metabolism; it could be required for the vegetative phase, fructification, or maturation of carpophores	10^{-6} M
Biotin (vit. B_7 or H)	Direct addition or from materials containing this vitamin	Coenzyme of pyruvate carboxylase or enzymes for carboxylation reactions of fatty acids or aspartic acid; donor or acceptor of carbon dioxide	100 µg/L
Nicotinic acid (vit. B_3), pantothenic acid (vit. B_3), and *p*-amino-benzoic acid	Present in natural materials; direct addition when synthetic media are used	Coenzymes for diverse groups of enzymes	5 µg/L or less

Table 8 shows the minerals and vitamins that are essential for the proper growth and development of fungi as well as the average contents required in culture media. To produce a particular substance or for the fructification of a given species, the definition of the specific needs of a mushroom in terms of inductors or inhibitors (such as vitamins and minerals) for the synthesis of the desired substances should be established.

Generally, the materials that make up the substrates used in the development and growth of white rot fungi are heterogeneous materials. In addition to the carbon and nitrogen sources, these materials contain an important amount of associated minerals (such as metals) that can serve as enzyme inductors depending on the concentrations at which they are present. These metals can be added to the culture medium using mineral salts that improve the reactions of the oxidoreductases produced by white rot fungi, such as *G. lucidum*. Metals such as copper, iron, manganese, molybdenum, zinc, and nickel are essential for the metabolism of white rot fungi. However, if they are found at higher than necessary concentrations, they can be toxic, causing a decrease in the growth rates and morphological changes in the mycelium (Baldrian 2003). Notably, low concentrations of these essential metals (iron, zinc, molybdenum, copper, and manganese) are essential for the development of the ligninolytic enzymatic system in many basidiomycetes species (Chang and Miles 1999). For example, copper and manganese are directly involved in the degradation of lignin; manganese participates in the reaction cycle of the manganese peroxidase (MnP) enzyme, and copper serves as a cofactor at the catalytic center of the laccase enzyme (Baldrian 2003).

Copper in the form of $CuSO_4$ acts as an inducer of laccase enzymatic activity because at high concentrations, it increases the superoxide dismutase, which is the enzyme responsible for producing hydrogen peroxide that, in turn, increases the activity of peroxide-dependent enzymes. This compound can bind to the cell wall, specifically to chitin and chitosan, and it can act as an inducer of the laccase enzyme with its four copper atoms in the nucleus, which stimulate the oxidation of each of the substrates and result in two water molecules.

This induction is given by a regulation at the transcriptional level since the mRNA levels increase in the presence of copper (Moreno and Ospina 2008). Although copper is part of the chemical structure of the laccase, MnP can also be affected by the concentration of copper in the culture medium; different investigations have shown that Cu^{2+} induces MnP synthesis in *Stereum hirsutum* (Mouso, Papinutti, and Forchiassin 2003) and laccase and MnP in *Trametes troggi* (Levin, Forchiassin, and Ramos 2002) when $CuSO_4$ is added to the medium.

Manganese is needed for the cyclic catalytic mechanism of the MnP enzyme to be completed. During this cycle, Mn^{2+} is oxidized to Mn^{3+}, which acts as a primary agent in the ligninolysis process when it is associated with organic acids such as malic and oxalic acid. However, high concentrations of this metal lead to enzyme inhibition, because large amounts of hydrogen peroxide are produced and then they break the catalytic cycle of MnP, affecting the amount of fungal biomass obtained as well as the mycelial growth of the fungus (Moreno and Ospina 2008). Although manganese and copper are the two metals that are directly involved in lignin degradation because they are part of the catalytic cycles of these enzymes, other metals, such as zinc, also affect their titers. For example, the addition of zinc to the culture medium at low concentrations (0.006-0.18 μM) increased the activity of the MnP and LiP in *Phanerochaete chrysosporium* and simultaneously increased the solubilization and mineralization of lignin (Baldrian 2003).

Along with the nutritional requirements required by mushrooms for growth and development, mushrooms also have minimal requirements regarding the physical conditions of their substrates to ensure proper development.

Thus, the most important physical factors are temperature, light, humidity, aeration, and gravity. Other factors that can affect the growth of some fungi are hydrostatic pressure, viscosity, and mutagenic radiation (Carlile, Watkinson, and Gooday 2001, Chang and Miles 2004).

SOLID-STATE CULTIVATION OF *G. LUCIDUM*

G. lucidum is the mushroom with medicinal attributes that has the greatest dissemination in the world. For this reason, the synthetic solid-state culture of this fungus is steadily growing around the world. At present, one of the most commonly used commercial cultivation methods for this mushroom corresponds to the traditional method, in which stumps and small trunks of selected woods are used as substrates. Nevertheless, the development of modern technologies for producing these fungal species broadens the spectrum of technological knowledge about its cultivation; for example, the most industrialized method uses mixtures of selected agro-industrial waste in bags or bottles (Chang and Miles 2004, Stamets 2000). The production of *G. lucidum* can be divided into two parts. The first part involves several steps, from the selection of the strains through the production of the commercial spawn or production inoculum. The second part corresponds to the preparation and development of culture media or substrates, which starts with the selection of feedstocks, the development of the formulations, sterilization, the inoculation of substrates, and the growth and development of the fungus until fruiting. This section describes the stages of *G. lucidum* mushroom production for industrial purposes.

Production of *G. lucidum* Inoculum

The inoculum of *G. lucidum* is produced through several steps involving the selection, maintenance, and conservation of the strains, the massification of the mycelial biomass (primary spawn), the production of secondary spawn, and the production of tertiary spawn. These steps are described in the following sections.

Strain Selection, Maintenance, and Conservation

The most common culture media for the growth and development of macromycetes that are on the market are potato dextrose agar (PDA) and malt extract agar (MEA). These media provide fungi with the nutrients necessary for their growth and development during the vegetative phase. When special growth conditions are required for a particular fungal species, it is possible to supplement these media with yeast extracts, peptone, specific vitamins or amino acids. Thus, these supplements contribute to the vigorous formation of the mycelial mantle and to the stimulation of the mycelium growth rate (Stamets 2000). Similarly, these culture media can be prepared from basic materials (components) and supplemented with salts, and an agar-agar base is added to provide consistency to the culture medium (Lin and Lin 2014). Table 9 shows alternative culture media for the growth and development of macromycetes fungi in the vegetative phase, which can be used for the massification of *G. lucidum*. The pH of these culture media is between 5.5 and 6.5. The culture media are sterilized in an autoclave at 121°C and 15 psig for 15 min. Next, the culture medium is poured into Petri dishes in a laminar flow cabinet and allowed to solidify; then, the appropriate fungal tissue is inoculated which is always added in the center of the Petri dish. The tissue may be a piece of the mushroom fruiting body when the strain needs to be renewed or a square piece of mycelium measuring 0.5 cm on each side that is spread on the agar. Subsequently, the inoculated Petri dishes are placed in an incubator with conditions of 25°C, 40%-50% relative humidity, and penumbra.

The *G. lucidum* strains are obtained from wild environments with subsequent vegetative reproduction and adaptation in a medium that is selective for their growth. In addition, these strains can be obtained from specialized laboratories dedicated to the collection of wild strains and the commercialized adaption of the original mycelium. To obtain wild strains, the specimens are collected, transported on paraffined paper, stored in a refrigerator and grown to the point of vegetative reproduction. For the vegetative reproduction of a specimen, a longitudinal cut is made in the mushroom fruiting body; a small part of the inside of the basidiome is sectioned from the fleshier area that is far from the hymenium, the structure where the spores are generated and stored. These pieces are washed in chlorinated water with 1-2% sodium hypochlorite for a maximum of 5 min, and then the pieces are washed in sterile distilled water. The washed mushroom fragments are inoculated onto PDA, MEA or other culture medium indicated for fungal growth in slant test tubes. Subsequently, the inoculated tubes are incubated at 25°C under penumbra for 10-12 days or until the colonization of the slant agar surface.

Macromycete growers should select the methods needed to determine the quality of new and stored strains when differences are detected in their growth and development, such as a retraction of the mycelium, a decrease in the growth rate on culture medium or other signals that can be detected during the production of the fruiting bodies (Stamets 2000). In addition, the mycelium that is spread on the agar of the working strain is subjected to specific biochemical tests to determine its degradation capacity. In particular, semi-quantitative tests of the laccase and cellulase enzymatic activities are performed on *G. lucidum* strains in a Petri dish according to the methodology by Baldrian and Valaskova (2008), which has been applied and adapted by Montoya, Sánchez, and Levin (2014). For this purpose, the strains that produce the largest halo of cellulose degradation and the ones that produce the largest halo during the laccase degradation tests are selected. Once the results of the biochemical tests are obtained, the mycelium is massified to generate the amount of mycelial biomass required to produce the commercial spawn or final inoculum.

G. lucidum is a species of white rot fungi. Its high cellulolytic and lignin-degrading activities imply high growth rates. Therefore, the assessment of these enzymatic activities is related to the quality of the *G. lucidum* mushroom strains.

In previous studies, semi-quantitative and qualitative methods were developed to detect the lignocellulolytic enzymes in white rot fungi. For example, several of these techniques have been developed to determine the degradation capacity of the lignocellulosic polymers by different white rot fungi. The polymers of the lignocellulosic biomass are diverse, and their chemical structures and conformations depend on their material of origin. Therefore, qualitative methods are a powerful tool for the rapid search (screening) of enzymes that can degrade the lignocellulosic complex (Nazareno-Saparrat et al. 2000, Pointing 1999). These tests provide an accurate indication of the presence of these enzymes to help researchers successfully select fungal strains by the group of enzymes they produce.

This selection strategy decreases the amount of necessary quantitative tests, which are much more expensive and require more time than the enzyme assays. However, these methods have limitations, especially regarding accuracy, although they play an important role in strain selection processes (Laureano-Perez et al. 2005, Malmström 1982). Moreover, these tests are useful for selecting strains from a storage strain bank (strain repositories) and they provide information on the strain integrity before being used for their final purpose. In the following research cases, the authors of this book disclose some procedures that have been developed to determine the viability of *G. lucidum* strains and to massify this mushroom during its vegetative phase.

Table 9. Culture media for mycelial vegetative growth of macromycetes

Culture media	Materials	Amount (g)	Procedure
Potato, Dextrose, Yeast Agar (PDYP or PDYPA if peptone is added)	Potatoes	300	Cook potatoes in 3 L water at boiling for 1 h; filter and remove the amount of liquid necessary to make up one liter from the filtrate; sterilize in an autoclave with the other components at 121°C and 15 psig for 15 min
	Glucose	10	
	Agar-agar	20	
	Yeast extract	2	
	Peptone	1 (optional)	
Malt Extract, Yeast Agar (MYA or MYAP if peptone is added)	Barley malt sugar	20	The materials are mixed, and water is added to make up one liter of medium; sterilize the medium in an autoclave at 121°C and 15 psig for 15 min
	Agar-agar	20	
	Yeast extract	2	
	Peptone	1 (optional)	
Oatmeal, Malt, Yeast Enriched Agar (OMYA)	Instant oatmeal	80	The materials are mixed, and water is added to make up one liter of medium; sterilize the medium in an autoclave at 121°C and 15 psig for 15 min; this medium does not need to be filtered
	Agar-agar	20	
	Malt sugar	10	
	Yeast extract	2	
Cornmeal, Yeast, Glucose Agar (CMYA)	Cornmeal	10	The materials are mixed, and water is added to make up one liter of medium; sterilize the medium in autoclave at 121°C and 15 psig for 15 min
	Agar-agar	20	
	Glucose	5	
	Yeast extract	1	

Adapted from Stamets (2000).

Research Case 2. Determining the Viability of Ganoderma lucidum Strains

Protocol for Determining the Strain Viability

In the Bioprocess and Agro-industry Plant at the Universidad de Caldas (Manizales, Colombia), a protocol for performing viability tests on the *G. lucidum* strains purchased from strain banks located abroad has been developed and is now always applied. These tests enable researchers to determine the quality of the strains in terms of their growth abilities and rates and the expression of the enzyme characteristic of each fungus, among other features. The protocol that was developed to determine the viability of the strains in different macromycetes species (including *G. lucidum*) acquired from international strain banks is shown in Figure 8.

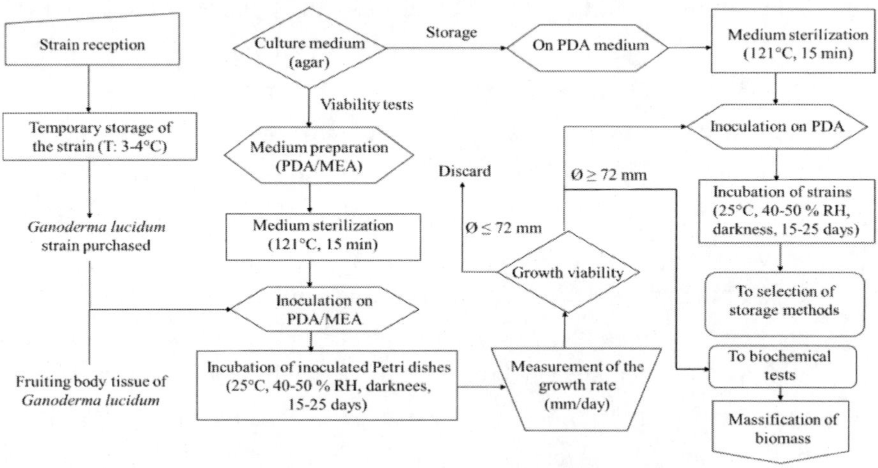

Figure 8. Protocol for determining the viability of *Ganoderma lucidum* strains from the Bioprocess and Agro-industry Plant at the Universidad de Caldas.

This protocol may also be applied when the strains are renewed from the extended tissue of a selected fruiting body; the procedure used to obtain this extended tissue is described briefly in the next subsection of this research case. The culture media used for the viability determination are PDA and MEA. The steps for performing the strain viability determination are as follows:

- Purchased strains or extended carpophore tissue from *G. lucidum* are kept in temporary storage under refrigeration at 3-4°C.
- Preparation of the PDA or MEA media onto which the strains will be inoculated. If both media are employed, the same number of Petri dishes with identical diameter measurements should be inoculated to compare the fungal growth on both culture media.
- Petri dishes must be washed and sterilized before use. Next, a Cartesian plane is drawn on the outer face of the base of each Petri dish with an indelible marker and ruler; all the Petri dishes should be numbered. At least five Petri dishes should be prepared per culture and per strain.
- Hot culture medium is poured inside a laminar flow cabinet and allowed to solidify.
- The culture medium in the Petri dishes is inoculated in the center of the plate, at the zero point of the Cartesian plane.
- The Petri dishes are incubated in an incubator or oven, in which the culture conditions may be maintained at 25°C with a dry environment (approximately 40% relative humidity).
- The growth diameter should be measured from the inoculation point. The value of this diameter must be determined on the two axes of the previously drawn Cartesian plane; that is, measurements are taken on the x-axis and y-axis and registered. This measurement should be taken every day until the first Petri dish reaches full colonization; this colonization corresponds to a diameter of 90 mm, but for this measurement, it is possible to accept an 89 mm-diameter mycelium in one of the axes of the Cartesian plane.
- When the first Petri dish presents a state of complete colonization, the incubation stage is considered finished. For viable and intact *G. lucidum* strains, the incubation time on PDA is approximately 10 days, and it is 12 days on MEA.
- By using the growth diameter of each plate and the time of colonization, the growth rate of the mycelium (in mm/day) is calculated. All the Petri dishes that have reached a diameter of

approximately 72 mm on one of the axes are selected; the other plates should be discarded.

- Petri dishes on which the mycelium has reached a colonization of 80% or more are subjected to biochemical tests consisting of qualitative or semi-quantitative laccase and cellulase (endoglucanases and exoglucanases) tests. These tests are described in Research Case 3.
- Once it is verified that the strains produce the enzymes essential for solid-state *G. lucidum* cultivation, one portion of the plates is employed for strain storage and maintenance, and the other one is used for biomass massification with the purpose of producing spawn and, subsequently, the fruiting bodies of the mushrooms.

Tissue Extension from G. lucidum Basidiomes

The renewal of a *G. lucidum* strain may be performed by considering the selection of a fruiting body that meets the requirements of the market. The carpophore is collected and wrapped in paraffined paper to be transported to the laboratory.

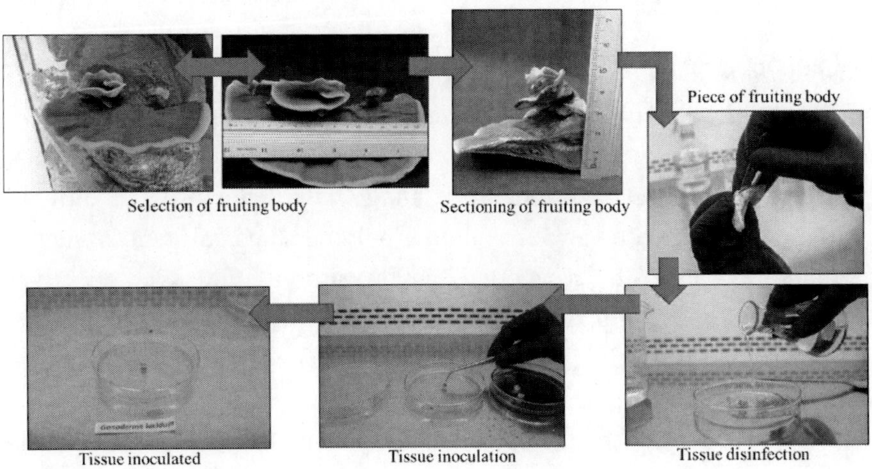

Figure 9. Overall procedure for renewing *Ganoderma lucidum* strains through tissue extension from their fruiting bodies in the Bioprocess and Agroindustry Plant at the Universidad de Caldas.

Prior to the collection of the basidiome, all the material necessary for this activity must be prepared. The required materials are: PDA or MEA medium that has been poured into the Petri dishes and solidified, a beaker containing sterile distilled water with 2% chlorine, two beakers containing sterile distilled water (400 mL each), a surgical use scalpel with a new blade, blotting paper, crystal flex paper, and the basic material needed to perform inoculations in the sterile medium. In addition, the laminar flow cabinet must be disinfected to perform what is known as a tissue extension. Figure 9 shows the procedure for renewing the working *G. lucidum* strains from the extension of tissue in the laboratory.

Research Case 3. Biochemical Tests for the Screening of Lignocellulolytic Enzymes as a Measure of the Lignocellulosic Material Degradation Capacity by Macromycete Strains

Among the white rot fungi considered to be powerful agents that produce lignocellulolytic enzymes, the following genera can be identified: *Auricularia*, *Schizophyllum*, *Lentinus*, *Polyporus*, *Pycnoporus*, and *Trametes*, among others (Montoya et al. 2010). Species of these genera, such as *Coriolus versicolor*, *Ganoderma lucidum*, *Ganoderma applanatum*, *Schizophyllum commune*, *Pycnoporus sanguineus*, *Pleurotus ostreatus*, *Grifola frondosa*, *Auricularia delicata*, *Trametes trogii*, and *Lentinula edodes*, are currently recognized for their high capacity for cellulose and lignin degradation. In addition, these fungi have been used in different processes to biodegrade various polluting polymers and substances, such as carbohydrates, aromatic compounds, hydrocarbons, and synthetic dyes. (Arora, Chander, and Gill 2002, Elisashvili et al. 2010, Litthauer et al. 2007).

In a previous study (Montoya, Sánchez, and Levin 2014), several biochemical tests were performed on ten species of macromycete fungi. These tests were performed to screen the lignocellulolytic enzymes of the evaluated species to assess their capacity to degrade the main polymers contained in the lignocellulosic materials that are usually employed as substrates for their growth and development.

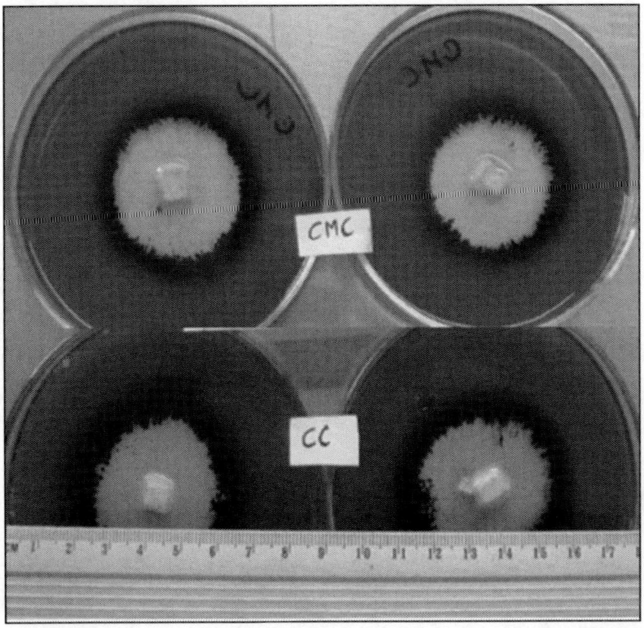

Figure 10. Formation of decolorization halos during the degradation test on two cellulose forms by *Ganoderma lucidum*. CMC: Carboxymethyl cellulose and CC: crystalline cellulose. Source: Montoya (2012).

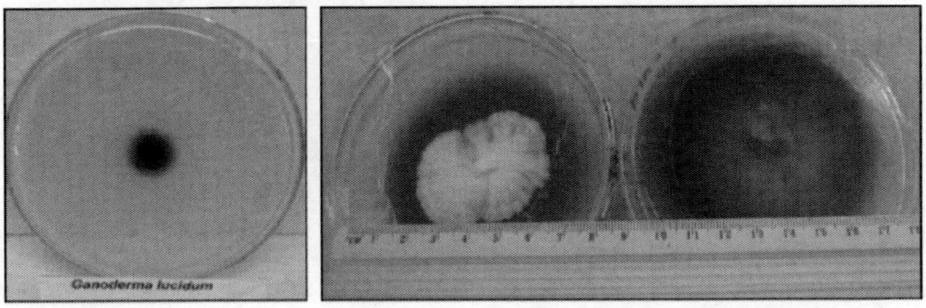

(a) (b)

Figure 11. Formation of green halos as an indicator of laccase production by *Ganoderma lucidum* mushrooms on a medium containing ABTS. (a) Pigmentation on the day of inoculation and (b) pigmentation of the medium at 14 days of incubation. Source: Montoya (2012).

The cellulose degradation capacity was determined by inoculating the mycelium of the strains screened on culture media containing 1.7% agar-agar and 0.1% carbon source. Carboxymethyl cellulose was used as a carbon source to determine the endoglucanase activity, while crystalline cellulose was used to determine the exoglucanase activity. The inoculated media were incubated for 5 days at 25°C. The colonies were then revealed by using 0.3% Congo red (an azoic dye) for 15 min, and the dye was subsequently rinsed with 1 N NaCl solution (Pointing 1999). Next, the resulting decolorization halos were observed and measured. These halos are an indicator of cellulose degradation since cellulose is complexed with Congo red; thus, the decolorization shows the absence of free cellulose in the medium (see Figure 10).

To estimate the laccase enzyme production in an organism, 0.035% (w/v) ABTS (2,2'azino-bis (3-ethylbenzo-thiazolin-6-sulfonic acid)) was used in a heterogeneous medium with the following composition (in g/L): glucose 10, K_2HPO_4 2, $MgSO_4{\cdot}7H_2O$ 0.5, yeast extract 0.2, $(NH_4)_2SO_4$ 0.5, $CaCl_2$ 0.1, $CuSO_4$ 0.0064, and agar-agar 22. The strain was inoculated into a representative number of Petri dishes with the necessary repetitions. The inoculated media were incubated for 12-14 days at 25°C. Subsequently, the green halo formation and the sizes of the colonies were determined in all cases (Pointing 1999). The formation of green halos as a consequence of the laccase enzyme action in *G. lucidum* can be observed in Figure 11. To estimate the production of lignin peroxidase, the heterocyclic dye Azure B was employed at a 50 μM concentration in the PDA medium. By using the same approach, the strains were inoculated into the media and incubated for 12-14 days at 25°C, and the decolorization halos were observed and measured.

Table 10 shows the average diameters of the decolorization halos observed during the cellulose degradation tests on the study species after they were revealed with Congo red. As shown in the data, *G. lucidum* exhibited high cellulose degradation ability, which implies a high potential to degrade and consume the different cellulosic materials contained in the given substrates.

Table 11 shows the results of the tests that were performed on the same ten species of macromycetes. These tests were used to determine the reaction capacity of these species on two specific substrates as indicators for the production of laccase and lignin peroxidase, two lignin-degradation enzymes characteristic of white rot fungi. From the results, it is evident that

G. lucidum has an enhanced ability to produce laccase and a moderate ability to produce lignin peroxidase.

Table 10. Average decolorization halos determined during cellulose degradation tests for ten white rot macromycetes species

Species	Halo diameter/mycelial growth (mm)
Auricularia delicata	8.0
Schizophyllum commune	6.5
Trametes trogii	26.0
*Coriolus versicolor**	65.0
Grifola frondosa	10.0
*Ganoderma lucidum**	59.0
Ganoderma applanatum	12.0
Pycnoporus sanguineus	27.0
*Lentinus edodes**	62.0
Pleurotus ostreatus	31.0

* Species exhibiting statistically significant differences related to the other species. Source: Montoya, Sánchez, and Levin (2014).

Table 11. ABTS oxidation or Azure B decolorization abilities of ten macromycete white rot species

Species	ABTS oxidation	Azure B decolorization
Lentinula edodes	++	-
Auricularia delicata	+	-
Grifola frondosa	+	+
Schizophyllum commune	-	-
Ganoderma applanatum	+	+
Ganoderma lucidum	++	+
*Coriolus versicolor**	++	++
*Trametes trogii**	++	++
Pycnoporus sanguineus	+	-
Pleurotus ostreatus	+	-

* Species exhibiting statistically significant differences in relation to the other species. Source: Montoya, Sánchez, and Levin (2014).

Research Case 4. Protocols for the Maintenance and Conservation of G. lucidum Strains

In the Bioprocess and Agro-industry Plant at the University of Caldas, several protocols for the maintenance and conservation of the working and research strains of different macromycetes have been developed. In particular, three methodologies are being used for the maintenance and conservation of the *G. lucidum* strains. The first methodology consists of placing pieces of mycelium that were previously grown on PDA into Eppendorf tubes with sterile distilled water; the tubes are closed with paraffin and stored at room temperature (17°C). This conservation method has shown viabilities of up to 90% during two years by making successive passages to fresh medium for growth tests and the evaluation of the main enzymatic activities. After passing these tests, the strains have been used for solid-state cultivation every two months.

The second methodology consists of the preservation of a piece of fungal mycelium spread on PDA in Eppendorf tubes containing 10% glycerol that are frozen at -20°C. This technique has been found to yield viabilities of up to 80% for three years; the resultant strains pass the successive growth tests and the evaluation of the main enzymatic activities. After passing these tests, the resultant strains have been used for solid-state cultivation every two months. The third methodology consists of the lyophilization of the mycelial biomass, which is stored in a freezer at -20°C. This procedure has been found to produce viabilities of up to 80%; the resultant strains pass the successive growth tests and evaluation of the main enzymatic activities. After passing these tests, the strains have been used for solid-state cultivation every two months.

When applying these three methodologies to *G. lucidum*, the most relevant characteristics are always monitored to evaluate and select the working strains with the best features. By using these methodologies, certain signs, such as depression of the mycelium and precipitation accelerated by tissue death due to an increase in diseases caused by polluting organisms or by the loss of the ability to produce necessary enzymes, are indicative of a need to renew the strains.

Massification of Mycelial Biomass

Massification involves successive passes of the mushroom mycelium onto PDA or MEA media using the already tested mycelium that has passed the viability and biochemical tests as original material, although it is possible to employ a spore isolation of the fungus. For this purpose, a square piece of mycelium measuring 0.5 cm on the side is inoculated onto agar-based medium with subsequent incubation at 25°C and under penumbra until complete colonization. As a result of the massification, a sufficient amount of mycelial biomass (called primary spawn) is obtained in the Petri dishes to produce the secondary and tertiary spawn required for commercial mushroom cultivation.

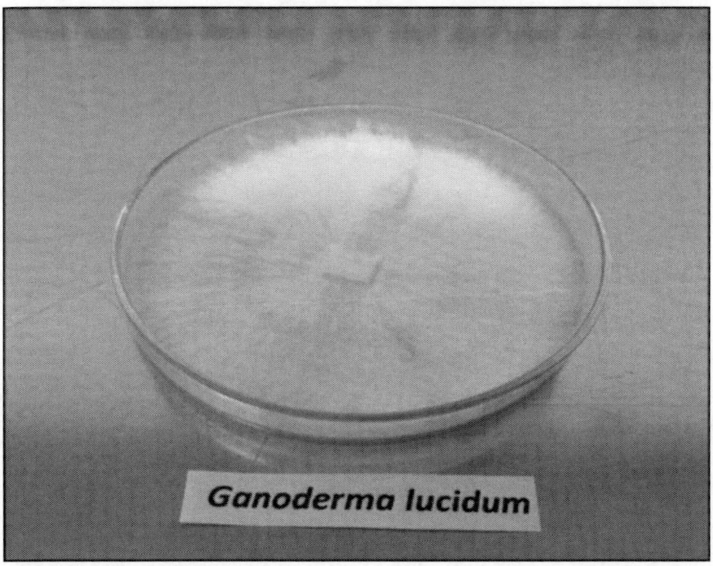

Figure 12. Mycelium of *Ganoderma lucidum* when extended on PDA.

For reishi mushrooms, massification consists of nothing more than reproducing the mycelial biomass successively obtained from the well-developed and fresh mycelium of *G. lucidum*, as presented in Figure 12. In the following research case, the procedures for *G. lucidum* massification are described briefly with respect to the protocol used in the Bioprocess and Agro-industry Plant.

Research Case 5. Massification of G. lucidum Biomass

The general scheme for the massification of the *G. lucidum* mycelial biomass is depicted in Figure 13. The procedures illustrated in this figure are also applied to the massification of *Grifola frondosa* strains. The starting point in this scheme is the strain that is purchased from a strain bank, which is called the lineage strain. This strain is tested for growth viability, as described above and in the scheme shown in Figure 8. The biochemical tests are subsequently performed, which were described above as well (see Research Case 3).

The massification of the *G. lucidum* mycelial biomass begins once the strain has passed the viability and biochemical tests. It is worth noting that each inoculation is equivalent to a new generation, as indicated in Figure 13. To produce mycelial biomass, it is necessary to calculate how many 90 mm Petri dishes containing culture medium and colonized by the working strain are required, considering that one of these Petri dishes can be divided into 24 to 30 pieces (see Figure 14). It is recommended that the pieces around the inoculation point of each plate should not be used as a new inoculum because the older mycelium is located in the center of the Petri dish.

Thus, the radial growth of the mycelium implies that the one at the periphery is the youngest and is usually lighter in color. These particularities are relevant during this stage of the process since, if the species is better known, the procedures for massifying the biomass are defined with greater certainty.

For example, the mycelium of *G. lucidum* is difficult to cut with a scalpel since it is separated from the agar at the time of sectioning because the mycelium is entangled, dense, and hard. During the increase of the biomass, the researcher seeks to ensure that all the mycelium in use exhibits the same quality features as the original strain, such as radial growth from the center of the Petri dish and the formation of concentric circles of growth with some yellowish pigmentation. In fact, these circles frequently demarcate natural changes in the age of the mycelium for the *G. lucidum* species.

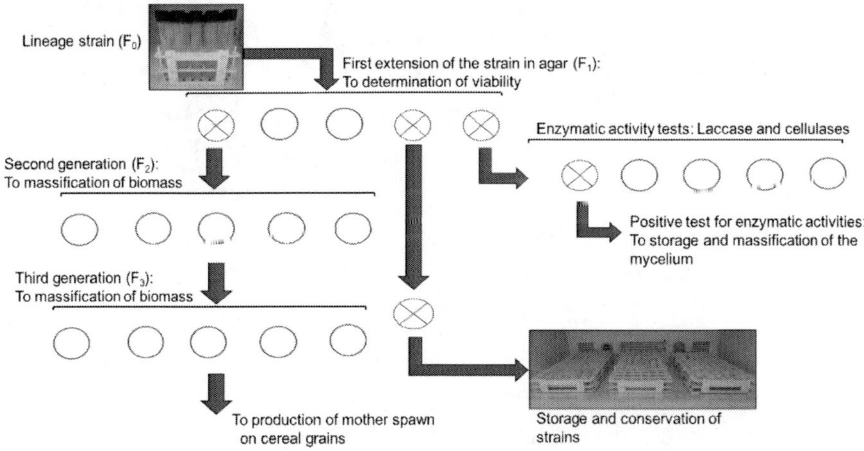

Figure 13. Scheme showing the massification of mycelial biomass from *Ganoderma lucidum* and *Grifola frondosa* strains in the Bioprocess and Agro-industry Plant at the Universidad de Caldas.

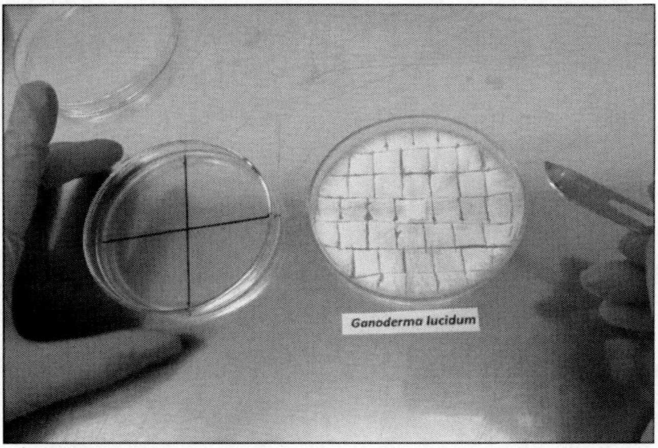

Figure 14. Division of reishi mycelium in a plate with colonized medium.

Production of Spawn

The production of commercial spawn (or production inoculum) from cultivable macromycetes such as *G. lucidum* is accomplished through the exponential expansion of the mycelial biomass on cereal grains or agro-industrial waste. The massified mycelium is induced to a state of vigor at which it is fully expressed, allowing for its adaptation to different types of lignocellulosic materials. When whole grain is used as a substrate, each grain becomes a mycelial capsule, which is a platform from which the mycelium can jump to the surrounding extension, i.e., to the complete the colonization of the grain. In general, all the cereal grains are potentially usable for spawn production. However, each species of macromycetes exhibits preferences for one cereal or another. Consequently, during the vegetative phase, each fungal species is provided the specific favorable conditions for its growth and development according to the cereal grain that is selected for this purpose (Stamets 2000).

The process of making commercial *G. lucidum* spawn is performed in two stages once the appropriate strain has been selected and massified as described in the previous sections. First, the mycelium is inoculated into grain so that the mushroom mycelium can be adapted for its subsequent cultivation in lignocellulosic substrates. The resulting product (extended mycelium in cereal grains) is called secondary spawn or mother spawn. Cereal grains are used whole to activate the production of lignocellulolytic enzymes by the fungus because they contain high molecular weight polysaccharides (mostly starch but also cellulose and hemicellulose), proteins, and minerals. During the second stage, the tertiary spawn of *G. lucidum* is produced using the secondary spawn as an inoculum at 3% to 5% (wet basis) of the culture medium, which is often also formulated on grains or different lignocellulosic materials. The production of this commercial spawn occurs in as many successive passes as required by the process and depends on both the availability of the initial mycelial tissue (primary spawn) and the amount of inoculum required for the final production of carpophores (Chang and Miles 2004, Stamets 1993, 2000).

It is worth highlighting that in the scientific literature and technical information, the primary spawn is also known as the agar inoculum, the secondary spawn (mother spawn) is also called the grain spawn or first-generation spawn, and the tertiary spawn is simply named the spawn (or second-generation spawn or production inoculum), which is the product that is employed to inoculate the solid substrates during the commercial production of the mushrooms.

Research Case 6. Development of G. lucidum Spawn Production

In the Bioprocess and Agro-industry Plant, the protocols for producing the *G. lucidum* spawn have been developed. These protocols were developed by considering the demand for the commercial spawn of this species in Colombia as well as the need to produce fruiting bodies of this mushroom at larger scales than that of the laboratory. The culture medium for spawn production may be hydrated cereal grains with up to 35-45% moisture or mixtures of lignocellulosic materials; formulations F1 and F8, which are described in Table 14, involve an adjustment of the mixture moisture to 60%.

The selection of the cereal grain for this purpose was performed by accounting for several factors, such as year-round grain availability in the market, competitive grain prices, and the high growth rates of the fungus on the grain. For example, *G. lucidum* exhibits an approximate growth rate of 1 cm/day on red sorghum and wheat. The mycelial growth of filamentous fungi occurs radially in all directions from the inoculation point in the substrate to which they are inoculated. The production of *G. lucidum* spawn is completed via two well-defined stages once the mycelium is massified: production of the grain spawn and the production of the commercial spawn or inoculum.

Production of Mother Spawn

The overall scheme for producing secondary *G.* lucidum spawn is shown in Figure 15, and the main process steps for producing the grain spawn of this fungus are illustrated in Figure 16.

The culture medium is called a "substrate" in this context. To propagate the mycelium in cereal grains, wheat or red sorghum grains are selected. The grain is washed with clean water to remove impurities, broken grains and powders. The washed grain is then hydrated in a kettle or by immersion in hot water, until its moisture reaches between 35% and 45%. Next, the grain is cooled to room temperature and packed in bioriented polypropylene bags in units of 200 grams.

The packed grain is sterilized at 121°C for 1 h with subsequent inoculation using four pieces of 0.5 cm mycelium on the side in the laminar flow cabinet. After the inoculation of the substrates, the polypropylene bags are conditioned with a PVC ring with a previously sterilized cotton plug to provide a filter during the incubation and storage of the grain spawn. The substrates are then incubated at 25°C, 50% relative humidity and under penumbra until reaching complete colonization. The incubation begins with the localized colonization in the grain from each inoculation point. Within approximately seven days, the secondary spawn reaches 60-70% of the colonization of the entire substrate. At this point, the solid culture medium is manually stirred or mixed to activate the mycelium growth, increase its growth rate and homogenize all the material; thus, the entire substrate completes its colonization at the same time. Moreover, this manual mixing at half the time of the incubation is useful for revealing possible contaminations that could have been introduced during the inoculation or during the first phase of the incubation. Finally, the grain spawn is ready when the substrate is completely colonized, acquires a white color, has a dense and compact mycelium, as shown in Figure 16.

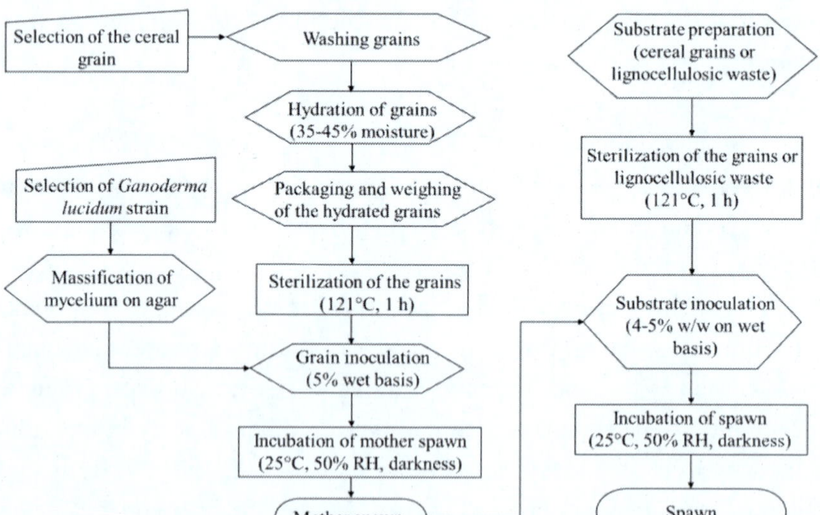

Figure 15. Scheme for producing the mother and the commercial spawn of *Ganoderma lucidum* in the Bioprocess and Agro-industry Plant. RH: Relative humidity.

Production of Second-Generation Spawn

The second-generation spawn of *G. lucidum* is produced according to the scheme depicted in Figure 15. For this purpose, the substrate is prepared using cereal grains or mixtures of agro-industrial waste according to the formulations for F1 or F8 presented in Table 14.

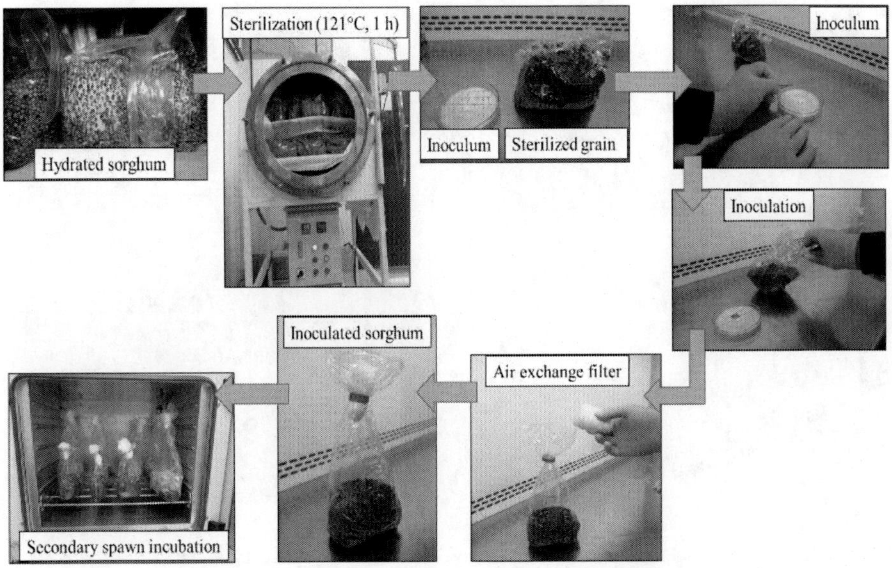

Figure 16. Process steps for producing the secondary spawn of *Ganoderma lucidum* in the Bioprocess and Agro-industry Plant.

Spawn production on cereal grains begins with the washing of the grain using clean water to remove impurities, broken grains, leaves, and powders. After the grains are washed, they are hydrated in boiling water until reaching a moisture level between 35% and 45%. Then, the grains are cooled to room temperature and packed in 1,000-g bioriented polypropylene bags. The substrate that is packed in the bags is sterilized for 1 h. Once the sterilized substrate has cooled, it is inoculated inside the laminar flow cabinet. The secondary spawn is used as inoculum at 3-5% (w/w) on a wet substrate basis, which corresponds to 30-50 g inoculum for each 1,000-g bag containing sterilized grain.

The inoculum is distributed homogeneously throughout the material. The bags containing inoculated material are provided with a PVC ring and a sterilized cotton plug, which is used as a filter for air exchange (see Figure 17).

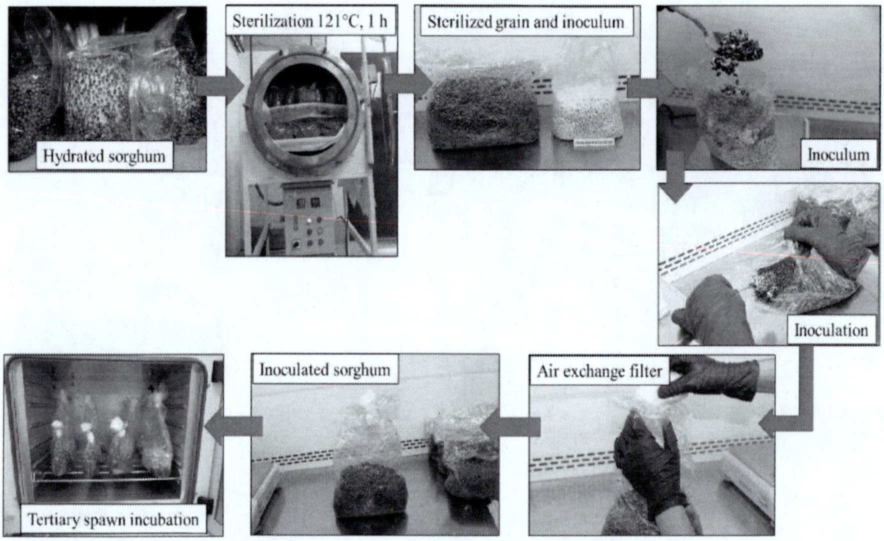

Figure 17. Process steps for producing the tertiary spawn of *Ganoderma lucidum* on grain in the Bioprocess and Agro-industry Plant.

The bags are taken to an incubation room at 25°C, 50% relative humidity, under penumbra, and with air exchange for 15 min per day. The incubation of the material begins with medium colonization from each inoculation point, that is, each cereal grain is colonized by the mycelium of the fungus. At the seventh day of colonization, the bags are manually mixed to activate the fungal mycelium, homogenize its growth, and perform contamination tests (if any). After mixing, the bags are left standing until 14 days of incubation, after which time the mycelium has completely colonized the substrate and turned white, thick, and dense. The incubation period of the second-generation spawn is 14 days for the complete colonization and thickening of the mycelium. The resulting spawn is stored under refrigeration at a temperature between 3°C and 4°C (maximum) for one month.

Production of *G. lucidum* Fruiting Bodies

G. lucidum is a fungus that naturally grows in subtropical and tropical zones, but it is rarely found in forests, which is why the cultivation of this species has been a subject of research and development for a long time. The artificial cultivation of reishi mushrooms was successfully achieved in the 1970s, and since the 1980s, the industrial production of this fungus has developed rapidly.

Large-scale *G. lucidum* cultivation was initiated on wood logs, and then formulations based on wood sawdust supplemented with nitrogen and mineral sources were developed. These solid mixtures were packaged in bags, sterilized and used for growth and development under controlled conditions (Chang and Miles 2004, Chen 1999, Hsu 1994, Mizuno et al. 1996).

Table 12. Substrate formulations proposed by Chang and Miles (2004)

Materials	F1	F2	F3	F4	F5
Sawdust	78.0	-	-	70.0	-
Bagasse	-	75.0		-	-
Cottonseed hull	-	-	88.0	-	-
Corn cob powder	-	-	-	14.0	78.0
Wheat/rice bran	20.0	22.0	10.0	14.0	20.0
Sucrose	-	1.0	1.0	-	-
$CaSO_4$	1.0	1.0	1.0	1.0	1.0
Soybean powder	1.0	1.0	-	-	-
Cereal straw ash	-	-	-	1.0	1.0

The culture media (substrates) for producing fruiting bodies of *G. lucidum* in indoor culture systems are formulated according to the availability of materials from the geographical areas in which the cultivation is expected to develop.

For instance, different formulations of culture media proposed by Chang and Miles (2004) are shown in Table 12. However, Stamets (2000) proposed a mixture of hardwood (alder or oak) sawdust and shavings enriched with 2.5 mL/L molasses solution. With these examples, it is possible to infer that in practically every geographical area, there are lignocellulosic materials that could be used for the cultivation of *G. lucidum*.

In the Bioprocess and Agro-industry Plant, different protocols for solid-state fermentation have been developed to produce fruiting bodies not only for *G. lucidum* but also for other white rot fungi with commercial and research purposes. Next, the procedures developed for these mushrooms are presented by using two research cases from the Laboratory of Macromycetes at the abovementioned plant.

Research Case 7. Polysaccharide Production by Submerged and Solid-State Cultures from Several Medicinal Higher Basidiomycetes

In a previous study (Montoya, Sánchez, and Levin 2013), the ability of ten species of white rot fungi (including *G. lucidum*) to produce extra and intracellular polysaccharides was investigated. Although the purpose of this study was to search for polysaccharides from fruiting bodies obtained by solid-state culture and from mycelium obtained by submerged fermentation in flasks, small-scale solid-state cultivation protocols were successfully developed for eight of the ten selected species.

In this way, the environmental factors required for producing the fruiting bodies of the study macromycetes were defined. The analyzed fungi were *Grifola frondosa*, *Trametes versicolor*, *Ganoderma applanatum*, *G. lucidum*, *Schyzophyllum commune*, *Lentinus edodes*, *Pleurotus ostreatus*, and *Trametes trogii*. This study was performed at the Universidad de Caldas from 2009 until 2012.

The culture medium used for these small-scale cultivation runs corresponded to the F8 formulation in Table 14. For *G. lucidum*, the initial substrate moisture was adjusted to 60%. The substrates were packed in 1-kg bioriented polypropylene bags and sterilized for 2 h using the Tyndall method, i.e., by heating the materials that were being sterilized to 121°C for 15 min in an autoclave. The results corresponded to the stages of growth and development of the eight species of macromycetes, which are described in Table 13.

Table 13. Basidiome development through synthetic "bag-log" cultivation for several white rot fungi and environmental conditions

Species	Spawn run (mycelial growth)	Primordial formation	Fruiting development	Carpophore morphological properties obtained during this study	Cultivation method adapted from:
L. edodes	25-28°C, darkness, 14.5% O_2, 5.5% CO_2, 60% RH, 50 days	Thermal shock at 8°C for 24 h, 14.5% O_2, 5.5% CO_2, darkness; 7 days at 18-22°C, 800 lux, 80% RH	18% O_2, 2.5% CO_2, 18-20°C, 800 lux, 80-85% RH, 7 days	Pileus 8-10 cm diameter, scaly, brown, with well-formed lamellae beige, stipe 5 cm long and 1 cm in diameter	Chang and Miles (2004)
S. commune	25-30°C, darkness, 60% RH, 20 days	18-20°C, 18% O_2, 2.5% CO_2, 700-1000 lux, 80-85% RH, 5 days	18-20°C, 18% O_2, 2.5% CO_2, 700-1000 lux, 80-85% RH, 7 days	Fan-shaped pileus grayish-white, with white lamellae, without stem, in clusters	Method developed in this study
T. trogii	25-30°C, darkness, 11% CO_2, 9.5% O_2, 60% RH, 30 days	18-20°C, 18% O_2, 2.5% CO_2, 400-500 lux, 70-80% RH, 25 days	18-20°C, 18% O_2, 2.5% CO_2, 400-500 lux, 70-80% RH, 7 days	Shelf-type pileus whitish yellow to honey, corrugated and continues as vertical blinds, all around the substrate block.	Method developed in this study
C. versicolor	25-30°C, 17.2% O_2, 7.3% CO_2 darkness, 60% RH, 30 days	18-20°C, 18% O_2, 2.5% CO_2, 500-800 lux, 70-80% RH, 20 days	18-20°C, 18% O_2, 2.5% CO_2, 500-800 lux, 70-80% RH, 14 days	Shelf-type pileus 4-6 cm wide showing typical concentric zones of different colors light tan, yellow to brown, individuals around the side of the substrate block	Method developed in this study
G. lucidum	27-30°C, darkness, 11% CO2, 9.5% O2, 60% RH, 35 days	18-20°C, 18% O2, 2.5% CO2, 500-700 lux, 70-75% RH, 40-50 days	18-20°C, 18% O2, 2.5% CO2, 500-700 lux, 70-80% RH, 15 days	Pileus 8-12 cm diameter, shelf-type, light yellow to red, with stipe 1-3 cm long, especially clustered at the top of the substrate block	Chang and Miles (2004)

Table 13 (Continued)

Species	Spawn run (mycelial growth)	Primordial formation	Fruiting development	Carpophore morphological properties obtained during this study	Cultivation method adapted from:
G. applanatum	27-30°C, darkness, 8% CO_2, 12.5% O_2, 60% RH, 25 days	18-20°C, 18% O_2, 2.5% CO_2, 300-500 lux, 70% RH, 45-60 days	18-20°C, 18% O_2, 2.5% CO_2, 300-500 lux, 70-80% RH, 15 days	Pileus 4-7 cm diameter, shelf-type, beige to dark brown, significant stipe, formed at the top and around the substrate block	Method developed in this study
G. frondosa	25°C, darkness, 14.5% O_2, 5,5% CO_2, 60% RH, 75 days	Thermal shock at 10°C for 24 hs 16-18°C, 19.5% O_2, 0.7% CO_2, 100-500 lux, 70-80% RH, 7 days	16-18°C, 19,5% O_2, 0,7% CO_2, 100-500 lux, 70-80% RH, 7 days	Pileus elongated, brown-dark gray to light beige, mature fruiting clusters with overlapping petals (caps and lateral stems) extending outward, resembling a flower cluster.	Montoya, Varón, and Levin (2008)
P. ostreatus	27-30°C, darkness, 9% CO_2, 11.5% O_2, 60% RH, 15 days	18-20°C, 18% O_2, 2.5% CO_2, 700-1000 lux, 90% RH, 5 days	18-20°C, 18% O_2, 2.5% CO_2, 700-1000 lux, 90-95% RH, 7 days	Pileus 7-10 cm diameter, oyster-shaped, purplish to light beige, stipe 1-3 cm, gregarious.	Rodríguez and Jaramillo (2005), Montoya, Restrepo, and Tabares (2009)

RH: Relative humidity. Source: Montoya, Sánchez, and Levin (2013).

For the solid-state culture of *G. lucidum*, an adaptation was made to the methodology used by Chang and Miles (2004). It is worth noting that the objective of this study was to develop the culture conditions of *G. lucidum* and other mushroom species to obtain quality carpophores as raw materials for intrapolysaccharide extraction and not to increase the production yields. Thus, the biological efficiency (*BE*) obtained for *G. lucidum* was only 8% on a dry basis in relation to the mass of dry substrate at the time of inoculation; the biological efficiency was determined by applying the equation (1). Nevertheless, the development of the solid-state culture of several species of white rot fungi at different scales began with this research. In particular, the solid-state cultivation of *G. lucidum* at the small and pilot scales was initiated using agro-industrial and agricultural waste to develop a process to produce this mushroom with a competitive *BE*.

Research Case 8. Solid-State Cultivation of G. lucidum at the Pilot Scale

Solid-state cultivation for the efficient production of *G. lucidum* fruiting bodies was developed ranging from small scale to pilot scale production for commercial purposes in the Bioprocess and Agro-industry Plant at the Universidad de Caldas. The cultivation protocols were defined by accounting for the nutritional requirements of the acquired reishi strains as well as the processes and environmental conditions necessary to obtain good-quality carpophores with competitive biological efficiency for the market. Colombia is a country with a great deal of agricultural activity, so it has large amounts of agricultural and agroindustrial waste that can be used for the cultivation of this mushroom. The formulations for which a better *BE* and fruiting body quality were obtained for this species are described in Table 14.

Table 14. Formulations of culture media (substrates) to produce *G. lucidum* fruiting bodies in the bioprocess and agro-industry plant at the Universidad de Caldas (in % dry basis)

Materials	F1	F2	F3	F4	F5	F6	F7	F8
Hardwood sawdust	73	50	60	48	50	40	30	73
Hardwood chips	-	-	-	30	-	31	30	-
Coffee husks	-	17	20	-	23	-	11	-
Cocoa husks	-	10	-	-	-	-	10	-
Corn bran	23	20	17	20	25	25	15	23
Sugar	2.0	1.0	1.0	-	-	2.0	1.0	2.0
Calcium carbonate	1.0	2.0	2.0	2.0	2.0	2.0	2.0	2.0
Calcium sulfate	1.0	-	-	-	-	-	1.0	-
Total	100%	100%	100%	100%	100%	100%	100%	100%

The formulations proposed for the production of *G. lucidum* fruiting bodies must be tested before being used on a large scale because the materials can be affected by their origin and storage. In addition, the carbon/nitrogen (C/N) ratios of all the formulations of substrates intended for *G. lucidum* cultivation must be determined as well. This parameter directly influences the growth and development of this fungus. Several C/N ratios of the formulations have been tested in the search for the best *BE* and fruiting body quality. During this process, it was determined that *G. lucidum* does not present significant variations in its EB nor in the quality of its fruiting bodies when the culture media (substrates) are formulated at a C/N ratio ranging between 40 and 80.

It has also been shown that adding selected salts at low concentrations also influences the growth and development of the fungus and the increase in the *BE*. For example, when adding copper and manganese salts to the formulations at concentrations no greater than 0.1% (w/w) on a dry basis, positive *BE* results can be obtained. The experiments performed at the Bioprocess and Agro-industry Plant showed that when salt percentages higher than those indicated above were used, the inhibition of the growth and development of the *G. lucidum* mycelium was verified. In addition, when percentages lower than 0.03% of these salts were added in equivalent proportions, no changes were observed in the growth and development of the fungal mycelium nor was there any difference in the *BE* obtained in the culture media with and without the addition of these salts.

When the substrates for the *G. lucidum* fungus are formulated, the amount of water added to the substrates is calculated so that the moisture reaches values between 62% and 65% in the initial mixture. The amount of water in the substrates must be calculated by performing a simple mass balance, and it is necessary to maintain the initial moisture of the feedstocks that make up the substrate.

For their vital functions, fungi take water from the solid material on which they grow. Therefore, each fungal species should be studied for the moisture interval necessary for proper growth and development. In a solid-state culture of macromycetes, the amount of added water directly influences fungal growth and development. If the moisture content is above the necessary levels, the water occupies the intraparticle spaces, interrupting the flow of air through the solid matrix of the substrate. This situation produces anoxia in the fungal cells, ultimately causing their death. In contrast, when the moisture content is below the level needed for macromycete cultivation, dehydration occurs in the hyphae, which leads to mass loss, and when the substrates reach fructification, the resulting fruiting bodies are small and weak.

In Figure 18, the overall scheme for producing *G. lucidum* fruiting bodies is depicted. After the materials for the formulation of culture media are selected, the substrates are packed in bioriented polypropylene bags with a capacity of 3 kg each. The bags are then sterilized at 121°C for 2 h. Figure 19 shows the preparation of the substrate mixtures and the filling of the bags, and Figure 20 shows their sterilization. The bags are subsequently transferred to an inoculation room with a laminar flow cabin. Figure 21 shows the inoculation procedure of the 3-kg bags of substrate with *G. lucidum* spawn. For this procedure, 4-5% (w/w) spawn referred to the wet weight of the substrate.

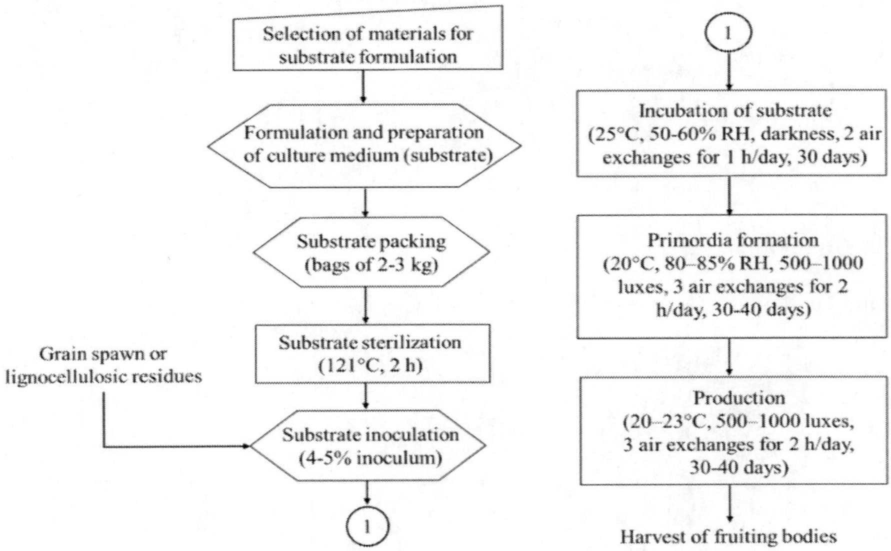

Figure 18. Overall scheme for producing *Ganoderma lucidum* fruiting bodies.

The inoculated substrates are housed in the culture rooms of the Experimental Unit of Macromycetes Cultivation of the Bioprocess and Agro-industry Plant. To determine the capacity of a cultivation room for *G. lucidum* cultivation, the area and height of the room must be considered, as the air head is an important factor to account for during mushroom growth and development. Thus, the growing rooms of this unit (see Figure 22) measure 2 m width × 3 m length × 3 m height (18 m^3) with an overall capacity of 200 kg substrate per room.

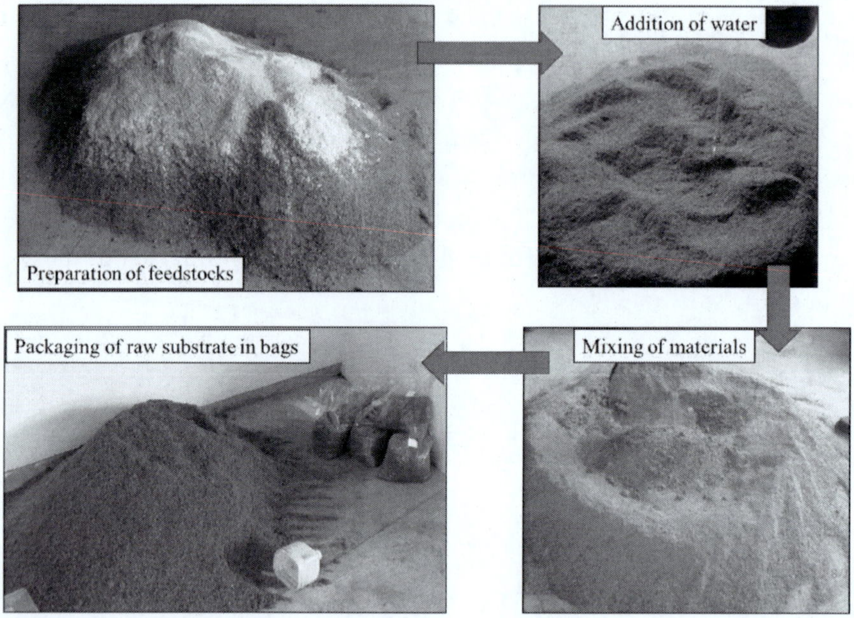

Figure 19. Preparing the bags and filling them with the formulated substrate.

Figure 20. Sterilization of the units with substrate packed for the cultivation of *Ganoderma lucidum* at the pilot and industrial scales.

Figure 21. Inoculation of substrates for cultivating *Ganoderma lucidum* at the pilot and industrial scales.

Figure 22. Producing *Ganoderma lucidum* in the Unit of Macromycetes Cultivation at the Bioprocess and Agro-industry Plant, Universidad de Caldas.

Table 15. Operating parameters for solid-state *G. lucidum* cultivation in a tropical climate (Manizales, Colombia) at the pilot scale for commercial purposes

Parameter	Incubation	Fructification		
		Primordia formation	Formation of young cap	Carpophore development until maturation
Temperature (°C)	25–30	18–23	20–23	18–23
Relative humidity (%)	95–100	90–95	90–95	90–95
Fresh air exchange	2 h/day	2–3 h/day	2– 3 h every 12 h	3–4 h every 12 h
Light intensity (lux)	Not required	700–1,000 during 12 h	700–1,000 during 12 h	700–1,000 during 12 h
Time required (days)	20–30	20–30	20–30	25–30
Overall time of the productive cycle		90 – 120 days		

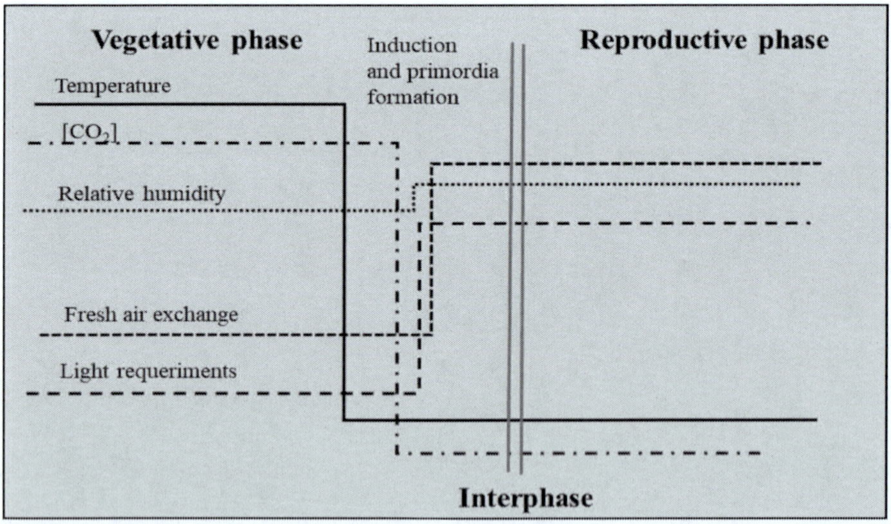

Figure 23. Transition of the environmental conditions between the vegetative and reproductive phases during the solid-state cultivation of *Ganoderma lucidum*. Adapted from Chang and Wasser (2017).

The operating parameters for growth and development required for the solid-state culture of *G. lucidum* in tropical climates are shown in Table 15. It is emphasized that the values of these parameters will change throughout the fungal development cycle. The environmental transition that must be guaranteed during the growth and development stages of the solid-state *G. lucidum* culture is shown in Figure 23.

In this scheme, it is possible to observe the changes that must be ensured in each of the main parameters during the stages of this culture. The formation of primordia becomes a substep of what is called the interface; that is, the major change in conditions between the vegetative phase (mycelial colonization of the substrate) and the reproductive phase (fructification).

These environmental changes become the most sensitive moments of the process; if they are not performed at the right time and with the necessary parameter modifications, the primordia are not formed, which leads to the loss of the culture.

The substrates are moved to the culture rooms to start the incubation process, and the conditions described in Table 15 are maintained. In this stage, which is the vegetative phase of the fungus, the mycelium colonizes the substrate, which acquires a whitish coloration. During this colonization, the fungal hypha attacks the substrate in the periphery and in the center. Once the mycelium has colonized all of the substrate, the formation of the mycelial coat begins, which corresponds to the thickening of the mycelium and a slight change of pigmentation from whitish to yellowish. The formation of the mycelial coat during the cultivation of *G. lucidum* is used to compact the blocks of substrate to generate the support necessary for the fruiting bodies.

After the formation of the mycelial coat, the formation of exudates and a yellow to brown pigmentation begins, which indicates that the blocks are ready for primordia induction. When most of the blocks (approximately 60%) in the culture room are pigmented, a change in the environment in the room is initiated to induce the primordia, as described in Table 15 and shown qualitatively in Figure 23. Once the environment changes in the culture room are completed, two horizontal slots are created around the perimeter of the blocks with a surgical scalpel to allow for the exit of the basidiomes, once they are formed. The fruiting bodies form whitish masses through these slots. After the formation of these amorphous masses, the growth of the carpophores begins, and they grow for 60 days until they reach full maturity.

Figure 24 shows the changes in the *G. lucidum* fungus from the colonization of the substrate to the mature fruiting body that is ready for harvest. The harvest of the carpophores is performed with scissors or a scalpel. A maximum of two harvests from each block is recommended. It should be clarified that if two harvests per block are expected in each room, then when the fruiting bodies of the first harvest are collected, the environment must be changed to induce the formation of the new primordia, which are formed from the points of the first harvest.

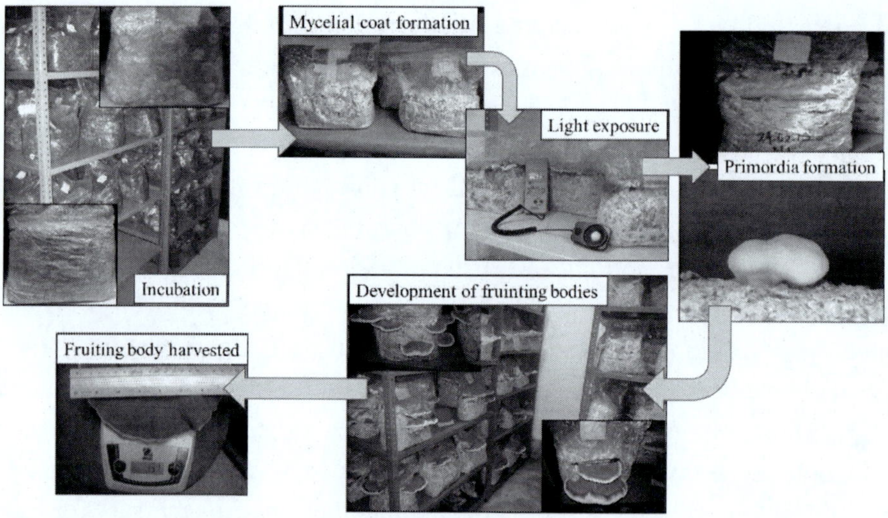

Figure 24. Stages of growth and development in *Ganoderma lucidum* from the vegetative phase to the reproductive phase.

Harvest and Post-Harvest of *G. lucidum*

The fruiting bodies of *G. lucidum* are harvested once the margin of the carpophores acquires the dark brown coloration observed in Figure 24. The *BE* reached in the Bioprocesses and Agro-industry Plant at the Universidad de Caldas is 20% (dry basis) on average in terms of the mass of the dry substrate at the time of planting according to equation (1). After the harvest, the fruiting bodies are chopped and subjected to a drying process at 40°C in a forced convection dryer. The pieces of the fruiting bodies are dehydrated for 24 h until a moisture of approximately 5-7% is reached.

Immediately after being dried, the fungi are ground in a centrifugal mill with a 200-mesh screen. The powdered dry *G. lucidum* is packed in 50-g jars or 100-g bags under vacuum for sale. Figure 25 shows the minimum processing procedure to which the *G. lucidum* fruiting bodies are subjected to obtain dry fungal powder for commercialization or for use as a raw material to extract bioactive substances or for the development of nutritional products with functional properties.

Figure 25. Minimum processing steps during the post-harvest treatment of *Ganoderma lucidum* mushrooms at the Bioprocess and Agro-industry Plant.

FINAL CONSIDERATIONS

In Colombia, for approximately two decades, small-scale trials have been conducted to support the cultivation of *Ganoderma lucidum* to reach the current technology level. In particular, in the Bioprocess and Agro-industry Plant at the Universidad de Caldas, the solid-state cultivation system for *G. lucidum* was developed with different available lignocellulosic materials, such as agricultural and agroindustrial wastes from around the country. In this regard, the proper protocols for maintaining and conserving the strains have already been defined, as have the development of all stages of pilot scale cultivation for commercial purposes and the industrial production of substrates inoculated with *G. lucidum* for industrial-scale production. The environmental conditions necessary for the efficient production of this fungus were also defined and established. Additionally, post-harvest treatment methods have been developed for the fruiting bodies.

All these developments are being offered to the national producers of the growing and dynamic sector of the medicinal mushroom industry. In fact, the Bioprocess and Agro-industry Plant has become a supplier of substrates that are inoculated with *G. lucidum* for several local and national companies. In addition, technical support and consultancy services for commercial producers are constantly being provided to support the technological development of this sector as a whole to increase its productivity and competitiveness.

REFERENCES

Arora, D. S., M. Chander, and P. K. Gill. 2002. "Involvement of lignin peroxidase, manganese peroxidase and laccase in degradation and selective ligninolysis of wheat straw." *International Biodeterioration & Biodegradation* 50:115-20.

Baldrian, P. 2003. "Interactions of heavy metals with white-rot fungi." *Enzyme and Microbial Technology* 32 (1):78-91.

Baldrian, P., and V. Valaskova. 2008. "Degradation of cellulose by basidiomycetous fungi." *FEMS Microbiology Review* 32:501-21.

Carlile, M. J., S. C. Watkinson, and G. W. Gooday. 2001. *The Fungi*. London: Academic Press.

Chang, S. T. 2009. *Training Manual on Mushrooms Cultivation*. Beijing: United Nations-Asian and Pacific Centre for Agricultural Engineering and Machinery (APCAEM).

Chang, S. T., and P. G. Miles. 1999. *Biología de las Setas. Fundamentos Básicos y Acontecimientos Actuales*. Bogotá: Instituto Zeri para Latinoámcrica. [*Biology of the Mushrooms. Basic Foundations and Current Events*. Bogotá: Zeri Institute for Latin America].

Chang, S. T., and P. G. Miles. 2004. *Mushrooms. Cultivation, Nutritional Value, Medicinal Effect, and Enviromental Impact*. London: CRC Press.

Chang, S. T., and S. P. Wasser. 2017. The Cultivation and Environmental Impact of Mushrooms. In *Oxford Research Encyclopedia of Environmental Science*. Oxford: Oxford University Press USA.

Chen, A. W. 1999. "Cultivation of the medicinal mushrooms *Ganoderma lucidum* (Curt.:Fr.) P. Karst. (Reishi) in North America." *International Journal of medicinal Mushrooms* 1:207-15.

Chen, W., and G. Miles. 1996. *Biomedical Research and the Application of Mushroom Nutriceuticals from Ganoderma lucidum in Mushroom Biology and Mushroom Products*. Hong Kong: Chinese University Press.

Elisashvili, V., E. Kachlishvili, T. Khardziani, and S. N. Agathos. 2010. "Effect of aromatic compounds on the production of laccase and manganese peroxidase by white-rot basidiomycetes." *Journal of Industrial Microbiology and Biotechnology* 37:1091-96.

Gottlieb, A. M., E. Ferrer, and J. E. Wrigth. 2000. "rDNA anlyses and an aid to the taxonomy of species of Ganoderma." *Mycological Research* 104:1033-45.

Hsu, Z. C. 1994. *New Technology for Cultivation of Ganoderma lucidum*. Liaoning, China: Chaoyang Edible Fungi Research Institute.

Hudson, H. 1986. *Fungal Biology*. Maryland, USA: Edward Arnolds Eds.

Laureano-Perez, L., F. Teymouri, H. Alizadeh, and B. E. Dale. 2005. "Understanding factors that limit enzymatic hydrolysis of biomass. Characterization of pretreated corn stover." *Applied Biochemistry and Biotechnology* 124 (1-3):1081-99.

Leskosek-Cukalovic, I., S. Despotovic, N. Lakic, M. Niksic, V. Nedovic, and V. Tesevic. 2010. "*Ganoderma lucidum* — Medical mushroom as a raw material for beer with enhanced functional properties." *Food Research International* 43 2262–69.

Levin, L., F. Forchiassin, and A. M. Ramos. 2002. "Copper induction of lignin-modifying enzymes in the white-rot fungus *Trametes trogii*." *Mycologia* 94 (3):377–83.

Lin, D., and Z. Lin. 2014. *Training Course on JUNCAO Technology for Latin America, Caribbean and South-Pacific Regions*. Edited by F. A. a. F. University, Fuzhou, China (Course documentation).

Litthauer, D., M. J. van Vuuren, A. van Tonder, and F. W. Wolfaardt. 2007. "Purification and kinetics of a thermostable laccase from *Pycnoporus sanguineus* (SCC 108)." *Enzyme and Microbial Technology* 40:563-68.

Malmström, B. G. 1982. "Enzymology of oxygen." *Annual Review of Biochemistry* 51 (1):21-59.

Mizuno, T., Y. Naoi, F. Mayuzumi, M. Ogino, and H. Okamoto. 1996. "Artificial cultivation of *Ganoderma lucidum* in Japan." In *Ganoderma lucidum* edited by T. Mizuno, 298. Seoul, Korea: II-Yang Pharmaceutical.

Montoya, S. 2012. "*Obtención de enzimas lignocelulolíticas y polisacáridos a partir de residuos lignocelulósicos del Departamento de Caldas empleando macromicetos de pudrición blanca por fermentación sumergida y fermentación en estado sólido.*" PhD Diss., Facultad de Ciencias Agropecuarias, Universidad de Caldas. ["*Production of lignocellulolytic enzymes and polysaccharides from lignocellulosic waste from Department of Caldas employing white rot macromycetes by submerged fermentation and solid-state fermentation.*" PhD Diss., Faculty of Agricultural Sciences, Universidad de Caldas].

Montoya, S., J. H. Gallego, A. Sucerquia, B. J. Peláez, O. Betancourt, and D. F. Arias. 2010. "Macromicetos observados en bosques del Departamento de Caldas: su influencia en el equilibrio y la conservación de la biodiversidad." *Boletín Científico Centro de Museos de Historia Natural* 14 (2):57-76. ["Macromycetes observed in the forests of the Department of Caldas: Their influence on the equilibrium and conservation of the biodiversity." *Scientific Bulletin Center of Museums of Natural History* 14(2):57-76].

Montoya, S., G. M. Restrepo, and L. A. Tabares. 2009. "Rendimientos en el cultivo de los hongos *Pleurotus ostreatus* y *Pleurotus pulmonarius* utilizando como sustratos diversos residuos agroindustrialcs." *Revista de Investigaciones Universidad Católica de Manizales* 14:16-26. ["Yields in the culture of the fungi *Pleurotus ostreatus* and *Pleurotus pulmonarius*, using different agro-industrial residues as substrates." *Research Journal Catholic University of Manizales* 14:16-26].

Montoya, S., O. Sánchez, and L. Levin. 2014. "Evaluación de actividades endoglucanasa, exoglucanasa, lacasa y lignina peroxidasa en diez hongos de pudrición blanca." *Biotecnología en el Sector Agropecuario y Agroindustrial* 12 (2):115-24. ["Evaluation of endoglucanase, exoglucanase, lacasse, and lignin peroxidase activities on ten white-rot fungi." *Biotechnology in the Agricultural and Agro-industrial Sector* 12(2):115-124].

Montoya, S., Ó. J. Sánchez, and L. Levin. 2013. "Polysaccharide production by submerged and solid-state cultures from several medicinal higher basidiomycetes." *International Journal of Medicinal Mushrooms* 15 (1):71-79.

Montoya, S., L. M. Varón, and L. Levin. 2008. "Effect of culture parameter on the production of the edible mushroom *Grifola frondosa* (maitake) in tropical weathers." *World Journal of Microbiology and Biotechnology* 24:1361-66.

Moreno, N., and X. A. Ospina. 2008. "*Evaluación de inductores metálicos y co-sustratos para la remoción de negro reactivo 5 empleando Pleurotus ostreatus inmovilizado en fique*." Trabajo de Pregrado, Facultad de Ciencias, Pontificia Universidad Javeriana. ["*Evaluation of metallic inductors and co-substrates for the removal of reactive black 5 using Pleurotus ostreatus immobilized in fique*." Undergrad. End Work, Faculty of Sciences, Pontifical Javeriana University].

Mouso, N., L. Papinutti, and F. Forchiassin. 2003. "Efecto combinado del cobre y pH inicial del medio de cultivo sobre la producción de lacasa y manganeso peroxidasa por *Stereum hirsutum* (Willd) Pers." *Revista*

Iberoamericana de Micología 20:176-78. ["Combined effect of copper and initial pH of the culture medium on production of laccase and managanese peroxidase by *Stereum hirsutum* (Willd) Pers." *Iberoamerican Journal of Mycology* 20:176-178.].

Nazareno-Saparrat, M. C., A. M. M. Bucsinszky, H. A. Tournier, M. N. Cabello, and A. M. Arambarri. 2000. "Extracellular ABTS-oxidizing activity of autochthonous fungal strains from Argentina in solid medium." *Revista Iberoamericana de Micología* 17:64-68.

Pointing, S. B. 1999. "Qualitative methods for the determination of lignocellulolytic enzyme production by tropical fungi." *Fungal Diversity* 2:17-33.

Rodríguez, N., and C. Jaramillo. 2005. "*Cultivo de hongos comestibles del género Pleurotus sobre residuos agrícolas de la zona cafetera.*" Boletín bot027. Chinchiná, Caldas, Colombia: Cenicafé. ["*Cultivation of edible mushrooms of the genus Pleurotus on agricultural waste of the coffee-growing zone.*" Bulletin bot027. Chinchiná, Caldas, Colombia: Cenicafé.].

Stamets, P. 1993. *Growing Gourmet and Medicinal Mushrooms*. Hong Kong: Ten Speed Press and Mycomedia.

Stamets, P. 2000. *Growing Gourmet and Medicinal Mushrooms*. Berkeley: Ten Speed Press.

Ying, J. Z., X. L. Mao, Q. M. Ma, Y. Z. Zong, and H. A. Wen. 1987. *Icons of Medicinal Fungi from China*. Beijing: Science Press.

Chapter 3

RESEARCH TRENDS ON *GANODERMA LUCIDUM* CULTIVATION AND ITS BIOLOGICALLY ACTIVE COMPOUNDS

INTRODUCTION

Currently, *Ganoderma lucidum* is one of the most in-demand medicinal mushrooms in the world. For this reason, different cultivation methods have been developed to produce mushrooms of better quality and with differentiated chemical compositions (Han, An, and Yuan 2005, Liu and Zhang 2007). For the artificial solid-state cultivation of Reishi mushrooms, several production alternatives have been generated and investigated to increase the biological efficiency of this culture and the quality of the resulting carpophores as well as to reduce the number of necessary production cycles, among other factors. Different and highly varied research efforts have been performed around the world to attain higher levels of efficiency and quality during the commercial production of *G. lucidum*, undoubtedly the medicinal mushroom with the greatest importance on the planet. One of the most important ongoing research and development trends is the application of JUNCAO technology for the large-scale production of Reishi fruiting bodies in several countries, especially in China. Moreover, different universities and research centers throughout the world are evaluating alternative feedstocks for Reishi cultivation as well as new, innovative methods to obtain *G. lucidum* mushrooms with increased concentrations of biologically active metabolites. For example, the influences of light at specific wavelengths are being assessed to enhance the biological efficiency and medicinal properties of Reishi mushrooms.

The fruiting bodies of *G. lucidum* contain polysaccharides, enzymes, glycoproteins, and other bioactive compounds such as triterpenoids, steroids, and fatty acids (Fukuzawa et al. 2008, Miura et al. 2003, Zhu and Lin 2005). Numerous biological activities associated with *G. lucidum* have been investigated, including anticancer activities, longevity effects, antiviral activities, and immunomodulatory properties, among others. Despite the knowledge accumulated about these activities, many issues and research challenges must be addressed to elucidate the complex mechanisms by which Reishi metabolites exhibit these important biological activities. In this way, new pharmacological applications may be developed to improve the health of the population and to support the treatment of serious diseases such as cancer.

This chapter covers some of the main research trends related to the development of new procedures for Reishi cultivation as well as the potential biologically active compounds that can be obtained from *G. lucidum* mushrooms.

G. LUCIDUM CULTURE USING ALTERNATIVE FEEDSTOCKS

The artificial cultivation of *G. lucidum* on a commercial scale has been reported since after World War II, specifically since the 1950s (Chang and Miles 2004). All the developments in the solid-state cultivation of this mushroom have focused on the utilization of sawdust from non-aromatic hardwoods as the basis of substrate formulations. However, due to the increase in the global market for *G. lucidum* because of its important medicinal attributes, the need to seek new cost-efficient and environmentally friendly alternatives for Reishi production has arisen. Specifically, the use of wild grasses to formulate substrates to produce *G. lucidum* fruiting bodies is a suitable alternative to meet these two requirements. In addition, other timber sources have been tested as alternative raw materials for Reishi production. These topics are described in the following subsections.

Solid-State Cultivation of Reishi by Using Grasses

Before the 1980s, forest resources were used as the main sources of feedstock for Reishi cultivation, including sawdust and logs. These resources were used for most mushroom cultivation around the world. In fact, the forestry sector in some regions, such as the Chinese province of Fujian (a region with a long tradition of mushroom production), has expressed concerns about using logs from broadleaf trees for these purposes. Therefore, experiments on edible fungus cultivation with herbaceous plants started in 1983, and *Lentinus edodes* production with wild grasses succeeded for the first time in 1986. The term JUNCAO was proposed by Professor Zhanxi Lin, a researcher at the Fujian Agriculture and Forestry University (Fuzhou, China), for grasses that are suitable for edible fungi cultivation. The JUNCAO technology was defined as techniques that employ JUNCAO to cultivate edible and medicinal fungi and to produce mycoprotein forage by using the spent substrate. All these outcomes have been developed at the JUNCAO Research Institute at Fujian Agriculture and Forestry University since 1983 (Lin and Lin 2014).

JUNCAO technology arose from the need to look for substitute materials to produce various macromycete species at different scales. Pastures and wild grasses become an important alternative to the use of wood residues for mushroom cultivation at the commercial scale. Likewise, the use of various grass species as raw materials for the ecological production of fungi on an industrial scale is a sustainable alternative around the world. The main advantages of this technology are as follows (Lin and Lin 2014):

- Use of herbaceous plants free of contaminants for the solid-state organic cultivation of edible or medicinal macromycetes.
- JUNCAO technology is favorable for the conservation of water and soil, as well as the protection of the ecological environment (especially the forest) and the sustainable development of the mushroom industry.
- Integration of sustainable production processes for plant cropping, animal farming, and fungus cultivation.

Progress in JUNCAO technology development ranges from the selection of grass species to the development of cultivation techniques for mushroom production using these grasses. The grass species selected for cultivating edible and medicinal mushrooms are presented in Table 16. Species of edible and medicinal mushrooms that have been cultivated by using JUNCAO technology are listed in Table 17.

A total of 42 species of higher grasses suitable for the cultivation of these fungi are known, and they were selected between 1983 and 2005 using systematic selection methods. In addition, during the same period, 50 species of mushrooms cultivated with the JUNCAO methodology have been selected. This solid-state cultivation technology for producing edible and medicinal fungi emerged in China as an important alternative to wood residues by using different grass species for competitive biological efficiency (*BE*).

JUNCAO technology was applied during the search for new alternatives to produce *G. lucidum* efficiently on an industrial scale. In 1987, Professor Lin began developing Reishi production by using JUNCAO as a substrate at Fujian Agriculture and Forestry University (Lin and Lin 2014). Grass species such as *Dicranopteris dichotoma*, *Miscanthus floridulus*, *Neyraudia reynaudiana*, *Saccharum arundinaceum*, *Themeda gigantea*, *Pennisetum purpureum*, *P. sinese*, and *Arundo donax*, among others, were found to be suitable substitutes for conventionally used forestry materials to cultivate *G. lucidum*.

Table 16. Grass species suitable for the cultivation of edible and medicinal fungi

Phylum	Class	Family	Genus	Species
Angios-permae	Monocoty-ledoneae	Gramineae	*Pennisetum*	*P. purpureum schumach* *P. sinese* Roxb *P. americanum* × *P. purpureum*
			Phragmites	*P. communis* Trin
			Neyraudia	*N. reynaudiana* (Kunth) Keng
			Miscanthus	*M. floridulus* *M. sacchariflorus* *M. sinensis*
			Saccharum	*S. arundinaceum* *S. sinensis* Roxb *S. robustum*
			Themeda	*T. gigantea* (Cav.) Hack *T. caudata* (Nees) A. camus
			Paspalum	*P. wettsteinii* Itackel *P. dilatatum*
			Vetiveria	*V. zizaniodes*
			Sorghum	*S. propinquum* (Kunth) Hitche *S. sudanense* *S. vulgare*
			Cymbopogon	*C. citratus* (DC.) stapf
			Arundo	*A. donx* L.
			Zea	*Z. mays* L.
			Triticum	*T. aestivum* L.
			Oryza	*O. sativa* L.
			Achnatherum	*A. splendens*
			Arundinella	*A. hirta* *A. nepalensis*
			Spartina	*S. anglica* C.E. Hubb.
			Setaria	*S. anceps*
		Musaceae	*Musa*	*M. nana* Lour
		Pontederiacea	*Eichhornia*	*E. crassipes*
		Polygonum	*Rumex*	*R. acetosa* L.
	Dicotyle-doneae	Malvaceae	*Gossypium L.*	*G. hirsutum* L.
		Leguminosae	*Medicago*	*M. sativa* L.
		Leguminosae	*Stylosanthes*	*S. guianensis*
		Asteraceae	*Helianthus*	*H. annuus* L.
		Asteraceae	*Silphium*	*S. perfoliatum* L.
		Umbelliferae	*Ferula*	*F. sinkiangensis*
Pterido-phyta		Gleicheniaceae	*Dicranopteris*	*D. dichotoma* *D. ampla*

Source: Lin and Lin (2014).

Table 17. Fungal species suitable for cultivation with JUNCAO

Subclass	Order	Genus	Species
Homobasidiomy-cetidae	Agaricales	*Agaricus*	*A. bisporus* *A. blazei* Murill
		Coprinus	*C. comatus*
		Pleurotus	*P. ostreatus* *P. sapidus* *P. rhodophyllus* *P. pulmonarius* *P. citrinipileatus* *P. cystidiosus* *P. ferulae* *P. abalonus* *P. eryngii* *P. tuber-regium*
		Armillaria	*A. mellea*
			A. tabescens
		Lentinula	*L. edodes*
		Flamulina	*F. velutipes*
		Tricholoma	*T. giganteum*
			T. lobynsis
		Hypsizygus	*H. marmoreus*
		Oudemansiella	*O. radicata*
		Stropharia	*S. rugosoannulata*
		Pholiota	*P. nameko*
		Volvariella	*V. volvacea* *V. bomycina*
		Agrocybe	*A. aegerita* *A. cylindracea*
	Russulales	*Hericium*	*H. erinaceum*
	Phallales	*Dictyophora*	*D. indusiata* *D. rubrovalvata* *D. duplicata*
	Aphyllophorales (Polyporales)	*Wolfiporia*	*W. cocos*
		Ganoderma	*G. lucidum* *G. sinensi*
		Coriolus	*C. versicolor*
		Grifola	*G. frondosa* *G. albicans* Imaz.
Heterobasidiae	Auriculariales	*Auricularia*	*A. auricula* *A. cornea* *A. polytricha* *A. pelitata* *A. delicata* *A. mesenterica*
	Tremellales	*Tremella*	*T. fuciformis* *T. aurantialba* *T. cinnabarina*

Source: Modified from Lin and Lin (2014).

The results of these developments showed that *G. lucidum* can grow and develop very well on JUNCAO, which has advantages such as faster fruiting, fewer production cycles, high productivity, good carpophore quality, and high contents of bioactive substances in the mushrooms. In addition, producing macromycetes with this methodology allows for the sustainable production of Reishi mushrooms and the use of spent substrates for new processes. The general composition of Reishi fruiting bodies grown using JUNCAO technology is as follows (Lin and Lin 2014): 12.10% crude protein, 79.33% crude fiber, 2.63% ash, 1.98% nitrogen, 0.23% phosphorus, 0.50% potassium, 0.085% calcium, and 0.056% magnesium. In addition, the total polysaccharide contents found in *G. lucidum* cultivated by using JUNCAO technology is higher than that found in carpophores grown on wood residues.

The overall procedure for cultivating *G. lucidum* with JUNCAO technology begins with the production of the grass species (JUNCAO) that is selected to cultivate the Reishi mushroom. This grass is harvested and milled to formulate the substrates and thus initiate the *G. lucidum* cultivation process. During this process, the stages of spawn production for substrate inoculation within the overall production cycle of fruiting bodies, including the harvest, drying, and storage of the carpophores, can be observed as well.

The substrate formulations developed for *G. lucidum* production within the JUNCAO technology framework are shown in Table 18. On average, the harvest of Reishi fruiting bodies begins at 2.5 to 3 months after the formation of the primordia by using these formulations and ensuring an adequate and viable environment for carpophore production. However, it must be noted that different strains of *G. lucidum* have different growth and development periods. In terms of the *BE* (see equation 1), JUNCAO technology has reached values from 18% up to 20%, while the biological efficiencies attained from substrates formulated with the sawdust of several wood species have reached only 15-17% (Lin and Lin 2014). In addition, investigators led by Professor Lin have performed research and development on the chemical composition and biological activities of Reishi fruiting bodies and spores produced by this technology.

Table 18. Substrate formulations for *G. lucidum* cultivation obtained by using JUNCAO technology (on dry basis)

Material	1	2	3	4	5	6	7	8	9	10	11	12	13	14	15
Dicranopteris dichotoma	30	20	30	39	20	26	25		40	30	40	30			25
Neyraudia reynaudiana							58	83							28
Saccharum arundinaceum		58													
Phragmites communis						52									
Miscanthus floridulus	48			39	33					23					
Pennisetum purpureum									43				76	63	30
Pennisetum sinese			53		30										
Paspalum wettsteinii											42				
Spartina alterniflora										25					
Sorghum propinquum												48			
Coffee shells														20	
Wheat bran	20	20	15	20	15	20	15	15	15	20	16	20	22	15	15
Calcium carbonate	2	2	2	2	2	2	2	2	2	2	2	2	2	2	2

Source: Lin and Lin (2014).

JUNCAO technology integrates techniques for cultivating macromycetes such as *G. lucidum* with the production of mycoprotein in forages and fertilizers. The creators of this technology highlight the following topics as the most relevant for process integration using grasses (Lin and Lin 2014):

- JUNCAO technology includes water control and soil erosion as well as the control of soil sterility and desertification. JUNCAO species and planting techniques are different for different climates or areas, such as hillsides, eroded soils, and desert barriers, among others.
- Development of JUNCAO processing and conservation techniques.

- Cultivation of edible and medicinal mushrooms using JUNCAO as substrates, such as *G. lucidum*, *Lentinus edodes*, *Auricularia auricula*, *Hericium erinaceum*, etc.
- Production of mycoprotein forage using JUNCAO.
- Efficient utilization of spent substrate produced by growing mushrooms based on JUNCAO as fodder and biofertilizer.

An integral system for JUNCAO use has been designed with multiple purposes, namely the conservation of hydric resources, feedstocks for producing edible and medicinal macromycetes, integrating by-products from mushroom production with the agricultural and animal farming productive chains, forage for poultry and livestock products, etc. Lin and Lin (2014) claim that JUNCAO technology makes the integration between productive cycles involving plants, fungi, and animals possible, thus establishing a beneficial economic, social, and ecological combination through the mushroom industry as an example of sustainable production.

Use of Alternative Timber Feedstocks for Reishi Production

Roy et al. (2015) tested wood sawdust from new materials in Bangladesh to evaluate the solid-state cultivation of *G. lucidum* to produce fruiting bodies from unconventional raw materials other than oak sawdust.

Table 19 shows the list of woods used to formulate the substrates. The substrates based on this sawdust were supplemented with rice and wheat bran to grow *G. lucidum*, and the corresponding cultivation times were determined. Substrates prepared from *Swietenia mahagoni* supplemented with wheat bran and rice bran first supported the formation of primordia at 9 and 12 days, respectively. The substrates that included the sawdust of *Dipterocarpus turbinatus* supplemented with rice and wheat bran started fruiting at 19 days of incubation. However, substrates formulated with sawdust from *Tectona grandis*, *Gmelina arborea*, and *Michelia champaca* and supplemented with rice and wheat bran exhibited poor, slow mycelial development that never fruited. Considering the substrates that fructified, the highest *BE* was 7.6 for the formulations containing *Swietenia mahagoni* sawdust supplemented with wheat bran, and the lowest *BE* was 3.6 for the formulation containing *Dipterocarpus turbinatus* sawdust supplemented with rice bran.

Table 19. Plant source for sawdust used for substrate preparation

Local name	English name	Scientific name	Family
Mahogani	Mahogami	*Swietenia mahagoni*	Meliaceae
Gerjan	Gerjan	*Dipterocarpus turbinatus*	Dipterocarpaceae
Segun	Teak	*Tectona grandis*	Verbenaceae
Gamar	Beechwood	*Gmelina arborea*	Lamiaceae
Chapa	Teak chambul	*Michelia champaca*	Verbenaceae

NOVEL METHOD FOR SOLID-STATE CULTIVATION OF *G. LUCIDUM*

Different improvements in the *G. lucidum* production process and in new supplements that enhance the biological efficiency and quality of its fruiting bodies have been proposed in recent years. Efforts are being made for this purpose in the Bioprocess and Agro-industry Plant at the Universidad de Caldas. In particular, the following research case describes a novel method for the cultivation and production of Reishi fruiting bodies based on the utilization of blue light during solid-state culture.

Research Case 9. Effect of Blue Light on the Solid-State Cultivation of G. lucidum

In the Bioprocess and Agro-industry Plant, the research project "Evaluation of the production of *Ganoderma lucidum* and its anti-inflammatory activity under two conditions of solid-state culture: Blue light and germanium-132 supplementation" was developed, as funded by the Universidad de Caldas and the National University of Colombia, Manizales. In this research case, the results obtained by evaluating the effect of two photoperiods of blue light on the solid-state culture of *G. lucidum* are presented.

The substrate used for this research had the following composition (on dry basis): 50% oak sawdust, 26.95% coffee husks, 20% corn bran, 2% calcium carbonate, 1% sugar, and 0.05% manganese sulfate. The substrate units were formulated with 62% moisture and packed in 1-kg bioriented polypropylene bags with subsequent sterilization at 121°C for two h by the Tyndall method; then, the bags were inoculated by adding 5% w/w (wet basis) spawn to the substrate in a laminar flow cabinet. Subsequently, the inoculated substrates were taken to a culture room and divided into three batches.

Figure 26. Employing a LED photoreactor for cell growth during the solid-state cultivation of *Ganoderma lucidum*.

A LED (light-emitting diode) photoreactor for cell growth, model ACL-002 A1 (Actum, Colombia) with two independent compartments and the ability to program the photoperiod, was employed. This photoreactor was located in one of the culture rooms of the Bioprocess and Agro-industry Plant (see Figure 26). In each compartment of the photoreactor, six bags containing 1 kg of inoculated substrate each were placed and subjected to photostimulation during all stages of *G. lucidum* growth and development.

In one compartment, the photo stimuli were applied for 12 h per day, while in the other compartment, the photo stimuli were applied for 24 h per day. In addition, six bags of substrate inoculated with the mushroom were subjected to conventional growth and development conditions and used as the control (without light stimuli), as described by Montoya, Sánchez, and Levin (2013). The conditions in the photoreactor were as follows: luminous power between 0 and 151.5 lumens per square meter (lm/m^2) for 12-h photostimulation and 92.6-151.5 lm/m^2 for 24 h photostimulation (continuous photostimuli); light beam angle of 110°, with a 0.35 m distance from the bulb to the geometric center of the experimental units. The illuminance (the total luminous power incident on a surface per unit area) achieved with a 10-W blue LED light bulb on the surface was measured with a double range photometer traceable to the geometric center of the experimental units (Montoya, López, and Segura 2018).

Table 20. Culture conditions for the growth and development of *Ganoderma lucidum* mushrooms under the stimulation of blue light

Culture conditions	Incubation			Production		
	Photoperiod 12 h	Photoperiod 24 h	Control	Photoperiod 12 h	Photoperiod 24 h	Control
Temperature (°C)	21–27	21–27	21–27	18–20	18–20	18–20
Relative humidity (%)	60–65	60–65	60–65	75–85	75–85	75–85
Luminous power (lm/m^2)	0 (darkness) 92.6–151.5 blue light every 12 h	92.6–151.5 blue light (continuous)	92.6-151,5 white light (continuous)	0 (darkness) 92.6–151.5 blue light every 12 h	92,6-151,5 blue light (continuous)	92,6-151,5 white light (continuous)
Air exchange	Not required	Not required	Not required	60 min every 12 h	60 min every 12 h	60 min every 12 h
Average process duration (days)	11.66±1.52	9.66±0.57	15.66±1.15	35.33±1.53	34.33±1.53	45.00±1.00

Source: Montoya, López, and Segura (2018).

The culture conditions for the growth and development of *G. lucidum* according to the photobiological treatments studied during the incubation and production phases of the mushroom are presented in Table 20. The shortest incubation and production times were obtained for the 24-h photoperiod, with averages of 10 and 34 days, respectively. Thc maximum total production cycle of *G. lucidum* mushrooms (61 days) corresponded to the control, to which fluorescent white light was applied. In contrast, the substrates treated with blue light had fewer production cycles.

The values of the productivity parameters and the size of the resulting carpophores are shown in Table 21. The biological efficiencies (*BE*) were calculated according to equation (1), while the production rate (*PR*) corresponds to the ratio between the biological efficiency and the total number of days of the process (Vega and Franco 2013). The highest values for *BE* and *PR* occurred for the 24-h photoperiod: 28.04% and 0.64, respectively.

These results are explained by the larger amounts of fresh mushrooms obtained and the greater reduction in the cultivation time. In contrast, the lower values of *BE* and *PR* corresponded to the control treatment: 23.96% and 0.40, respectively. The total time for the solid culture of *G. lucidum* was reduced by 13 days and 16 days for the treatments with photoperiods of 12 and 24 h, respectively.

Table 21. Productivity parameters and size of *Ganoderma lucidum* fruiting bodies when grown under blue light

Treatment	Production cycle (days)	*BE* (%)	*PR*	Fresh mushroom mass (g)	Size of basidiomes first crop (cm)	
					Length	Width
Photoperiod 12 h	47±3	26.48±0.45	0.56±0.022	911.47±19.73	13.0±4.6	6.5±1.4
Photoperiod 24 h	44±2	28.04±0.43	0.64±0.008	964.85±14.06	13.6±2.8	7.0±1.3
Control	60.6±2.1	23.96±0.38	0.40±0.015	824.68±11.42	12.4±1.8	7.8±1.5

BE – biological efficiency; *PR* – production rate. Source: Montoya, López, and Segura (2018).

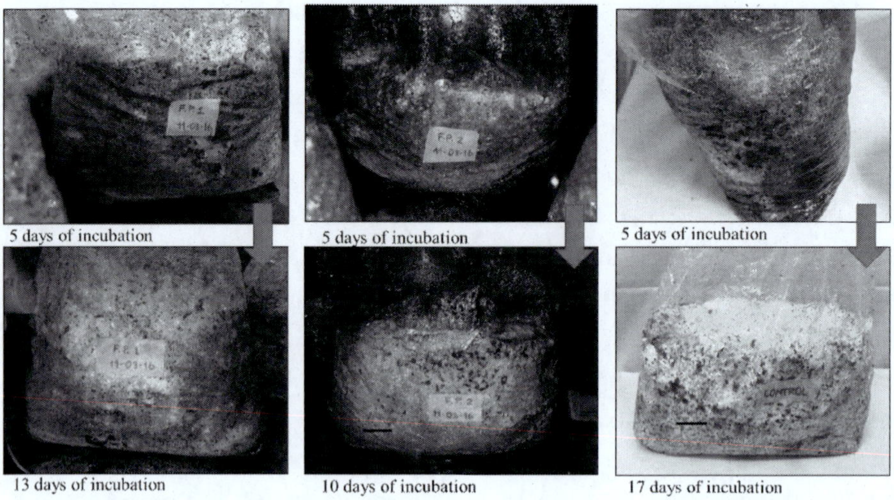

(a) (b) (c)

Figure 27. Substrate colonization by *Ganoderma lucidum* until the end of the incubation phase for three light treatments. (a) Blue light with 12-h photoperiod. (b) Blue light with 24-h photoperiod. (c) Control treatment with white light.

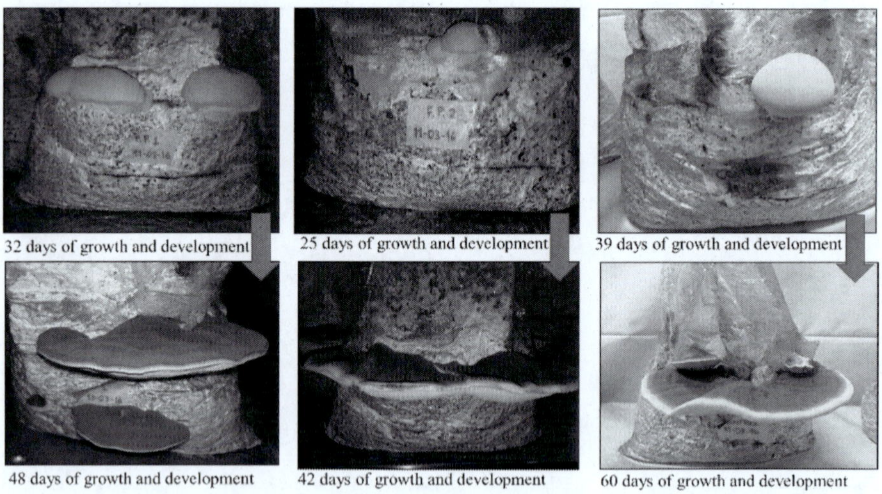

(a) (b) (c)

Figure 28. Growth and development of *Ganoderma lucidum* basidiomes for three light treatments. (a) Blue light with 12-h photoperiod. (b) Blue light with 24-h photoperiod. (c) Control treatment with white light.

Figure 27 shows the progress in the substrate colonization by *G. lucidum* mycelia for the light treatments evaluated here. Greater growth was observed in the substrates with blue-light stimulation relative to the control substrates. The lowest time in the incubation phase was obtained for the substrates that were continuously stimulated with blue light. For this treatment, the total incubation time was reduced down to 10 days, while the control treatment required 17 days to complete the colonization of the substrate.

Figure 28 shows the development of the *G. lucidum* basidiomes during the fructification phase for the three light treatments assessed here. Again, the shortest growth time until the moment of harvesting was obtained for the continuous treatment with blue light (42 days), which was 70% of the time required for complete carpophore formation up to the harvest point in the case of the control treatment (60 days).

The biotechnological process developed here included exposing the substrate units inoculated with *G. lucidum* to blue light, which was more efficient for producing fruiting bodies than the conventional process. Increases of 17% in the biological efficiency and 60% in the production rate were reached for the 24-h photoperiod treatment relative to the control treatment. In addition, a reduction in the cultivation time was obtained, leading to reduced process costs. An improvement in the quality of the fruiting bodies was also attained. The basidiomes obtained from 24-h photostimulation using blue light were those with the best appearances; they were 10 to 17.5 cm thick and fleshy, with an average weight of 68.21 g per specimen. Therefore, it is inferred that the stimulation with blue light during the solid-state cultivation process of *G. lucidum* has important positive effects on both the productivity and quality. These results show that there is a relationship between the productivity parameters and the light. In particular, specific light wavelengths (in this case, blue light) favor the mycelium growth and fructification of Reishi mushrooms. These outcomes may become the starting point for the industrial scale-up of solid-state *G. lucidum* cultivation as assisted by blue light.

BIOLOGICALLY ACTIVE COMPOUNDS FROM *G. LUCIDUM*

Ganoderma, which is called Lingzhi in China, was identified taxonomically as *Ganoderma lucidum* (Leyss, Ex Fr.) Karst (Chinzi) and *G. sinense* Zhao, Xu and Zhang (Zizhi) belonging to the Polyporaceae family. Lingzhi is part of the Pharmacopeia of China (National Pharmacopoeia Committee 2010). More than 200 *Ganoderma* species have been found in the world, of which 98 have been found in China, with *G. lucidum* and *G. sinense* being the most commonly used for medicinal purposes. These species have been cultivated commercially for the preparation of medicinal products since the 1970s (Dong and Han 2015). Currently, several types of products based on medicinal mushrooms such as *G. lucidum* are available on the market, as dry powder from the fruiting bodies, powdered and combined preparations from substrate, mycelium and primordia, biomass or extracts obtained by submerged fermentation, tablets or capsules from dehydrated fruiting bodies, spores, and other extracts (Wasser 2014). Different natural health products based on fungi such as the Reishi constitute a large market that can exceed 2.5 billion US dollars per year (Bidegain et al. 2014).

Many researchers have found that *G. lucidum* possesses approximately 400 different bioactive compounds, mainly triterpenoids, polysaccharides, proteins, nucleotides, steroids, fatty acids, and trace elements. Different studies have shown that Reishi mushrooms exhibit properties related to health improvements and the fight against diseases such as hypertension, hepatitis, diabetes, neurasthenia, and cancer. Of all the types of compounds found in the fruiting bodies of *G. lucidum*, triterpenoids and polysaccharides have been identified as the most relevant biologically active components (Dong and Han 2015, Jin et al. 2016).

Several terpenoids and their derivatives have been isolated from fungal species in the Polyporales and Ganodermatales orders and have been shown to be cytotoxic. At least 100 different triterpenoids from the fruiting bodies and mycelia of *G. lucidum* and *G. applanatum* have been identified. These compounds include ganoderic, ganoderenic, and lucidenic acids as well as ganoderals (Wasser and Weis 1999). However, Reishi mushrooms contain β-*D*-glucans along with glucuronoglucan, xyloglucan, unannoglucan, xylomannoglucan, and other active complexes of heteroglucans and proteins, according to the analyses performed on hot water extracts from this mushroom (Mizuno et al. 1995).

One of the most important medicinal attributes of the Reishi mushroom is its immunomodulatory activity. It has been proven that several *G. lucidum* substances (e.g., polysaccharides, triterpenoids, and proteins) have immunomodulatory effects, such as increased activity in T effector cells, natural killer (NK) cells, and macrophages. The β-glucans of *G. lucidum* appear to be one of the first biological response modifiers for which the cellular mechanism of action was initially defined at the receptor-specific level. Regarding this mechanism, the host cytotoxic defense function of β-glucans is specific for the target cells, and its action in promoting the host defense is based on the specificity of the antibody (Smith, Rowan, and Sullivan 2002). These polysaccharides are responsible for regulating the expression of cancer genes, and triterpenes are known as suppressors of cancer cell proliferation as well as the cancer cell cycle (Ćilerdžić et al. 2014, Ofodile et al. 2005). This section highlights the antitumor and immunomodulatory activities of *G. lucidum* as well as other important pharmacological properties exhibited by this mushroom. The role played by the polysaccharides from the Reishi is crucial. The most important results of different research efforts to elucidate these properties are briefly discussed below. With this purpose, an overview on polysaccharides is presented in the next subsection.

Generalities about Polysaccharides

Polysaccharides are condensed polymers from a large number of monosaccharides. The predominant monosaccharide in polysaccharides is *D*-glucose, although monosaccharides such as *D*-fructose, *D*- and *L*-galactose, *D*-xylose, and *D*-arabinose are also common. Monosaccharide derivatives are also found among the hydrolysis products of natural polysaccharides, including *D*-glucosamine, *D*-galactosamine, *D*-glucuronic acid, N-acetylneuraminic acid, and N-acetylmuramic acid. The monosaccharides are linked by glycosidic bonds between the glycosyl residue of the hemiacetal and the hydroxyl residue of another monosaccharide. Polysaccharides may be linear or branched, and with the exception of cyclic polysaccharides known as cycloamyloses or cyclodextrins, their chains have a nonreducing end (or ends) and a reducing end. In general, these biopolymers differ in the nature of their repeated monosaccharidic units, in the length of their chains, and in their degree of branching. The general formula of polysaccharides is $C_x(H_2O)$, where x is usually a number between 200 and 2500 (Aluko 2012, Izydorczyk 2005, Lehninger 1981, Zong, Cao, and Wang 2012).

Table 22. Primary natural homopolysaccharides

Type	Polysaccharide	Repeating unit (Glycosidic linkage /monosaccharide unit)
Linear	Amylose	α-(1→4)-Glc
	Cellulose	β-(1→4)-Glc
	Xylan	β-(1→4)-Xyl
	Inulin	β-(2→1)-Fru
	Levan	β-(2→6)-Fru
	Laminaran	β-(1→3)-Glc
	Chitin	β-(1→3)-GlcNAc
	β-glucan	β-(1→4, 1→3)-Glc
	Curdlan	β-(1→3)-Glc
Branched	Amylopectin	α-(1→4, 1→6)-Glc
	Dextran	α-(1→2, 1→3, 1→4, 1→6)-Glc
	Levan	α-(2→1, 2→6)-Fru
	Pullulan	α-(1→6)-Maltotriose
	Scleroglucan	β-(1→3, 1→6)-Glc
	Glycogen	α-(1→4, 1→6)-Glc
	Lentinan	β-(1→3, 1→6)-Glc
	Grifolan	β-(1→3, 1→6)-Glc
	Schizophyllan	β-(1→3, 1→6)-Glc

Remarks: Fru – fructose, Glc – glucose, NAc – N-acetylglucosamine.
Source: Sánchez, Montoya, and Vargas (2015).

The polysaccharides can be divided into two classes, depending on the number of different monosaccharides in their structure: homopolysaccharides (homoglycans), which contain a single type of monosaccharide, and heteropolysaccharides (heteroglycans), which contain two or more different types of monosaccharide units. The homopolysaccharides, which have linear or branched chains, are commonly named by the monosaccharidic unit that composes them; they have the same or different glycosidic bonds according to their configuration (α or β) and/or their position as those listed in Table 22. However, heteropolysaccharides can have different types and sequences of glycosidic bonds between their monosaccharide units, and there is great variety in these structures (Belitz and Grosch 1997, Izydorczyk 2005, Lehninger 1981).

The polysaccharides synthesized by fungi can be intrapolysaccharides (they are found inside the cells, where they exhibit their functionality) or exopolysaccharides (polysaccharides released into the medium where the fungi grow). The exopolysaccharides are generally glucans made up of glucose, which present differences in their structures, molecular weights, and composition, depending on the organism where they are found or the environment into which they are excreted. Thus, when they are released into a liquid medium, they can be recovered through different extraction processes. However, a large amount of polysaccharides is found as intrapolysaccharides that make up part of the structure of the carpophores or mycelium (Carlile and Watkinson 2000). Fungal polysaccharides show large differences in their solubility. In particular, the solubility of β-glucans in water increases when the degree of polymerization decreases. β-glucans are insoluble when the degree of polymerization is greater than 100.

Overview of the Bioactive Components of Fungi

Proteins are one of the main components of fungi, especially in the case of edible mushrooms. Mushrooms are considered a source of digestible proteins with a protein content above that of most vegetables, but lower than that of most meats and milk. The protein content can vary from 10-40% on a dry basis. Mushrooms contain all the essential amino acids but can be limited in amino acids containing sulfur, cystine and methionine (Chang

1991). In addition, mushrooms have a significant carbohydrate content. For example, fresh button mushrooms contain 3-21% carbohydrates and 3-35% fiber on a dry basis. Thus, a considerable proportion of mushroom carbohydrates is composed of polysaccharides that humans cannot easily digest, but that function essentially as dietary fiber, and thus the caloric value of most mushrooms is low (Smith, Rowan, and Sullivan 2002).

Numerous fungi are characterized as a source of different bioactive polysaccharides (see Table 23). Most of them are homoglycans and some are heteroglycans, although they can also be found in combination with proteins, forming peptidoglycans or polysaccharide-protein complexes (proteoglycans). The fungal cellular wall consists of homoglycans such as chitin, β-(1,3)-glucan, β(1,6)-glucan, α(1,3)-glucan, heteroglycans, and other compounds such as mannoproteins.

β-glucans are found in the cellular wall of microorganisms such as bacteria. In fungi, they are located in the intermediate layer of the cellular wall, adjacent to the plasma membrane; their function is to maintain the rigidity and shape of the cell. Some of the best known fungal β-glucans in terms of pharmacological applications are lentinan, schizophyllan, and krestin (PSK), which are mostly β-(1→3, 1→6)-glucans (Sánchez, Montoya, and Vargas 2015) whose overall structure is depicted in Figure 61. Lentinan, which is obtained from the fungus *Lentinus edodes*, has a triple helix structure and a molecular weight between 400 to 800 kDa. Schizophyllan is obtained from the fungus *Schizophyllum commune*, and it has a triple helix structure and a molecular weight of approximately 450 kDa. PSK is a proteoglycan made up of 25-38% protein residues and β-(1,4)-glucan with side chains of β-(1,6)-glucopyranosides, which are obtained from *Coriolus versicolor*, and it has a molecular weight of 94 kDa (Becerra-Jiménez et al. 2011, El Enshasy and Hatti-Kaul 2013, Mantovani et al. 2008, Zhang et al. 2013).

Numerous research efforts have been conducted in the search for useful substances from fungi for treating cancer due to the incidence of this disease around the world. Fungal metabolites can be used to inhibit molecular targets in malignant cells to fight certain types of cancer. Fungal anticancer substances can be roughly divided into two groups: high and low molecular weight molecules. The main difference between these two groups is their mechanism of action. Most high molecular weight compounds are polysaccharides or polysaccharides bound to proteins (Kidd 2000). It seems that these compounds are able to interact with the immune system in a nonspecific way to regulate the host response (Lull, Wichers, and Savelkoul 2005). The low molecular weight group includes secondary compounds of low molecular weight (such as terpenoids), which can penetrate the cell membrane and act specifically in transduction pathways (Zaidman et al. 2005).

Table 23. Bioactive fungal polysaccharides

Bioactive compounds	Source	Health benefits	Reference
Heteroglucans and homoglucans	*Agaricus blazei, Ganoderma lucidum, Lentinus edodes, Grifola frondosa, Coriolus versicolor, Schizophyllum commune*	Anticancer and antitumor activities	El Enshasy and Hatti-Kaul (2013), Hirahara et al. (2012), Mantovani et al. (2008), Ramberg, Nelson, and Sinnott (2010), Wang et al. (2014), Wu et al. (2012), Yamanaka et al. (2012), Yue et al. (2012), Zhang et al. (2013), Ziliotto et al. (2009)
β-(1→3)-glucan with β-(1→6)-branching, β-(1→3) branched β-(1→2)-mannan, β-(1→3)-glucan, cordyglucan, lentinan, glucan, mannoglucan, galactomannan, grifolan, schizophyllan, acidic polysaccharides, heterogalactan, krestin, heteroglycan, scleroglucan (β-(1→6)-monoglucosyl-branched β-(1→3)-glucan), heteroglucans	*Agaricus blazei, Cordyceps sinensis, Cryptoporus volvatus, Ganoderma lucidum, Grifola frondosa, Hericium erinaceus, Inonotus obliquus, Lentinus edodes, Morchella esculenta, Phellinus linteus, Pleurotus ostreatus, Polystictus versicolor, Schizophyllum commune, Sclerotinia sclerotiorum, Tremella aurantialba,*	Immunomodulators. The immune-stimulating effect of β-glucan is probably associated with the activation of cytotoxic macrophages and T-helper and natural killer (NK) cells with the promotion of T lymphocyte differentiation and activation for the alternative complement pathway	El Enshasy and Hatti-Kaul (2013), Mantovani et al. (2008), Moradali et al. (2007)

Table 23 (Continued)

Bioactive compounds	Source	Health benefits	Reference
β-glucan and exopolysaccharides	*Agaricus blazei, Cordyceps sinensis, Lentinus edodes, Phellinus baumii, Tremella fuciformis*	Antidiabetic effect	De Silva et al. (2012)
Prepared derivatives of (1→3)-β-D-glucan	*Saccharomyces cerevisiae*	Antibacterial, antimutagenic, antioxidant, antitumor and immunostimulating activities	Kogan and Kocher (2007)
Pullulan: α-(1→6) linked polymer of maltotriose subunits	Synthesized by fermenting a liquefied starch, coconut by-products, beet molasses, agro-industrial waste with *Aureobasidium pullulans*. Chemical reactions.	Pullulan has been used for liver and tumor targeted delivery of drugs. Pullulan has an application of targeting drugs to liver and cancer cells	Prajapati, Jani, and Khanda (2013)
EPS2	Polysaccharide produced by the marine filamentous fungus *Keissleriella* sp. YS 4108	EPS2 exhibited profound free radical-scavenging activities	Laurienzo (2010)
Exopolysaccharides	Marine fungus *Penicillium* sp. F23-2	Three polysaccharides from *Penicillium* sp. F23-2 possessed good antioxidant properties, especially superoxide and hydroxyl radical scavenging abilities	Laurienzo (2010)

Source: Sánchez, Montoya, and Vargas (2015).

It is known that cancer and its treatment by conventional therapies such as chemotherapy and radiotherapy have adverse effects on the human immune system. Cancer immunology has become a rapidly growing field in basic cancer research. Currently, new ways to promote the activity of the host antitumor immune cells and to overcome the ability of cancer cells to evade immune surveillance are being sought (Cassileth 2000, Curt 1998). Immuno-stimulating agents such as fungal polysaccharides are possibly useful complements for conventional cancer treatments if they do not interfere with the ability of conventional treatments to kill tumor cells. Byerrum et al. (1957) were the first to publish the antitumor properties of fungal polysaccharides, which is why, since then, many researchers around the world have devoted themselves to studying the properties of these substances in conjunction with other metabolites contained in macromycetes or metabolites that are isolated from these fungi. The polysaccharides of the macromycetes activate the innate immune system and exert antitumor activity by accelerating the host defense mechanism. Hence, these polysaccharides activate effector cells, such as macrophages, T-lymphocytes, B-lymphocytes, cytotoxic T-lymphocytes (CTL), and natural killer cells to express cytokines such as TNF-α, IFN-c and IL-1β. Cytokines always possess antiproliferative activity, cause apoptosis and differentiation in tumor cells, and secrete products such as reactive nitrogen, oxygen intermediates and interleukins (Collins et al. 1991, Paterson 2006). Additionally, fungal polysaccharides generate activity in cancer cells, and this activity consists of a series of complex reactions that stimulate the innate and adaptive immune system (Chan, Chan, and Sze 2009).

Ganoderma is the most popular and intensely researched genus among the fungi and has the greatest medicinal attributes. Many of its species are known for their multiple medicinal activities. These activities are related to the production and contents of various metabolites, such as polysaccharides, proteins, terpenes, sterols, etc., which are produced and contained in the mycelia and fruiting bodies of the species in this genus. However, most reports and scientific literature refer almost exclusively to *G. lucidum* (Wasser et al. 2006). In particular, *G. lucidum* exhibits a high concentration of biologically active polysaccharides, usually at a higher level and number than in other natural resources (Smith, Rowan, and Sullivan 2002).

Immunomodulatory and Antitumor Properties of *G. lucidum*

Mushrooms in general, and *G. lucidum* in particular, contain important biologically active compounds that are relevant to human health called "mycochemicals", such as antimicrobial metabolites (antibacterial, antifungal), antiviral metabolites, cytotoxic substances, immunomodulators, antineoplasia substances, and compounds with miscellaneous biological activities.

Among these mycochemicals, polysaccharides play a crucial role, especially considering their immune-modulatory and antitumor properties, along with triterpenes/triterpenoids, proteins, peptides and amino acids, nucleosides, nucleotides and RNAs, alkaloids, vitamins, essential minerals, and fatty acids (Paterson 2006). Several biologically active compounds from *G. lucidum* and *G. frondosa* are presented in Table 24. In general, each one of the metabolite groups from these two mushroom species has been linked to specific antitumor activities. For example, polysaccharides help the host to withstand adverse biological stress and increase immunity against the development of cancer cells (Meng, Liang, and Luo 2016).

More than 100 types of polysaccharides have been isolated from *G. lucidum*. Some Reishi β-*D*-glucans have shown immunomodulatory and immune-stimulating activities (Gao et al. 2003). The structure of β-glucans plays a role in their biological activity, as do their size and solubility. Larger forms of β-glucans molecules derived from the cellular walls of mushrooms, especially in their native triple-helical form, or particles derived from β-glucans, appear to be more potent than smaller molecules (Bohn and BeMiller 1995).

In general, water-soluble β-glucans have been studied extensively and have been shown to have a variety of immunostimulatory (activate immune responses) or immunomodulatory (regulate immune functions) effects (Batbayar, Lee, and Kim 2012, Rice et al. 2005).

Table 24. Primary composition of *G. lucidum* and *G. frondosa* and related bioactive substances

Substances	*G. lucidum*	*G. frondosa*	Biological activities
Water	6.90%	17.40%	
Protein	26.40%	20.30%	Higher nutritional functions
Fat	4.5% stearic acid, palmitic acid, oleic acid, lignoceric acid, *n*-nonadecanoic acid, tetracosanol, hentriancotane and choline triacylglycerols	3.5% stearic acid, palmitic acid, lignoceric acid, n-nonadecanoic acid, tetracosanol, behenic acid, tetracosanol triacylglycerols	Oleic acid, an inhibitor of histamine release
Food Fiber	0.10%	0.10%	The dietary fiber in fungus could lower the cholesterol level and prevent atherosclerosis, constipation, and diabetes
Carbohydrates	43.1%, principally β-*D*-glucan (25%)	46.0% principally β-*D*-glucan (20%), of which 20% is the MD-fraction	Immunologic, anti-inflammatory and anticancer effects involving several pathways via the *Dectin-1* receptor (see above)
Vitamins (mg/100 g product)	Vit. B_1 3.5 mg, Vit. B_2 17 mg, Vit. B_6 0.7 mg, choline 1150 mg, niacin 62 mg, inositol 307 mg	Vit. B_1 3.8 mg, Vit. B_2 11 mg, choline 850 mg, niacin 73.5 mg, inositol 347 mg, betaine 5.4 mg	
Terpenes	Ganoderic acid (triterpenes)	Various sesquiterpenes	Anti-androgenetic, G1 cell cycle blocking, inhibition of cell adhesion and migration
Inorganic (mg/100 g product)	Calcium 832 mg, phosphorus 4150 mg, iron 83 mg, magnesium 1030 mg, natrium 375 mg, potassium 3590 mg	Calcium 820 mg, phosphorus 4550 mg, iron 86 mg, magnesium 930 mg, natrium 355 mg, potassium 3390 mg	Cyclo-octasulfur, a strong inhibitor of histamine release. Exceptional source of potassium
Others	Higher RNA level	Moderate RNA level	RNA that induces interferon production in human cells, which disrupts viral invasions
	Isoflavones	High level of isoflavone	Mimic the estrogen hormones in humans

Source: Rossi et al. (2018).

Numerous studies have shown that *G. lucidum* polysaccharides exert their antitumor activity in the host immune system indirectly instead of through direct cytotoxic effects. In general, there is evidence that β-*D*-glucans induce a biological response by binding to membrane complement receptor type three (CR3, integrin αMβ2, or CD11b/CD18) in immune effector cells such as macrophages (Chan, Chan, and Sze 2009, Gao, Chan, and Zhou 2004). In addition, it has been demonstrated that β-glucans can also help to override the normal resistance of tumor cells to the cytotoxic activation of phagocyte and natural killer (NK) cell CR3, allowing this important effector mechanism of the complement system to function against tumor cells (Ross et al. 1999).

G. lucidum has been recommended as a coadjuvant agent in cancer chemotherapy treatments. *G. lucidum* extracts are known to possess cytotoxic activity against several cancer cell lines, such as leukemia, lymphoma, multiple myeloma (Müller et al. 2006, Tomasi et al. 2004), PLC/PRF/5 human hepatoma, MDA-MB-231 and MCF-7 human breast cancer (Jiang et al. 2004), PC3 human prostate cancer, and human cervical tumor HeLa (Zhu et al. 2000). The antitumor activity of *G. lucidum* has been associated with the use of regular doses of the powder from its fruiting bodies or mycelia, and multiple tests performed directly with specific purified and isolated molecules have yielded excellent experimental results. However, the molecular mechanism responsible for the inhibitory effects of this mushroom on cancer cells has not been fully clarified (Mahajna et al. 2008).

The valuable antitumor properties of Reishi extracts have been researched in relation to prostate cancer. Prostate cancer is the most common male malignancy in all Western countries and is the third leading cause of death from cancer in men worldwide (Parkin et al. 2002). This type of cancer tends to be a disease of older men, with more than 70% of the diagnoses performed in men over 65 years old. Androgens play an important role in the proliferation, differentiation, maintenance, and function of the prostate, but they are also involved in the development and progression of prostate cancer. Androgen action is mediated by the androgen receptor (AR), which binds to the active androgen dihydrotestosterone (DHT), translocates into the nucleus and regulates androgen-responsive genes that have been implicated in the development of this cancer (Lee and Chang 2003, Zaidman et al. 2008, Zaidman et al. 2005). Zaidman et al. (2008) studied the anti-androgenic activity of organic *Coprinus comatus* and *G. lucidum* extracts. Dry mycelia from the two species were ground into fine powders and extracted with ethyl acetate, diethyl ether, or ethanol for 7 days. Subsequently, the solvents were removed under reduced pressure; ultimately, these extracts were stored in a stock solution that was prepared in dimethylsulfoxide and stored at -70°C. These extracts showed good results at selectively decreasing viability and inducing a G1 phase arrest in androgen-dependent LNCaP cells. In addition, these extracts showed the ability to decrease luciferase reporter activity in MDA-kb2 breast cancer cells induced by DHT and to reduce the secreted levels of AR-regulated prostate-specific antigen in DHT-induced LNCaP cells, and they were able to decrease the levels of AR protein in LNCaP and MDA-kb2 cell lines. These authors state that metabolites from these mushrooms are excellent candidates as AR modulators, so the results obtained in vitro showed the potential use of Reishi extracts in cancer therapeutics.

Considering the valuable and relevant biologically active properties of Reishi mushrooms, Hobbs (1995) summarized the main pharmacological effects of whole *G. lucidum* extracts that have been tested in vivo and in vitro (generally in the form of aqueous or ethanolic extracts), as follows:

- Analgesic
- Anti-allergic activity
- Bronchitis prevention, inducing the regeneration of the bronchial epithelium
- Anti-inflammatory
- Antibacterial, against *Staphylococcus* spp., *Streptococcus* spp., and *Bacillus pneumoniae* (perhaps due to increased immune system activity)

- Antioxidant, by eliminating hydroxyl free radicals
- Antitumor activity
- Antiviral effect, by inducing interferon production
- Lowers blood pressure
- Enhances bone marrow nucleated cell proliferation
- Cardiotonic action, lowering serum cholesterol levels with no effect on triglycerides, enhancing myocardial metabolism of hypoxic animals, and improving coronary artery hemodynamics
- Central depressant and peripheral anticholinergic actions on the autonomic nervous system to reduce the effects of caffeine and relax muscles
- Enhanced natural killer cell activity in vitro in mice
- Expectorant and antitussive properties demonstrated in mouse studies
- General immunopotentiation
- Anti-HIV activity in vitro and in vivo
- Improved adrenocortical function
- Increased production of interleukin-1 by murine peritoneal macrophages in vitro
- Increased production of interleukin-2 by murine splenocytes in vitro.

This author warns that most of these pharmacological effects were determined from the results of uncontrolled clinical or laboratory studies or reports, mostly from China. The key active constituents of the Reishi that could be responsible for the listed effects are primarily β-glucans and hetero-β-glucans (antitumor and immune-stimulating activities), the so-called Lingzhi-8 protein (anti-allergenic and immunomodulatory activities), ganodermic acid, and other triterpenoids (agents that reduce cholesterol and blood pressure, anti-allergenic agents). Since the time that this list was published, many of these properties have been demonstrated, as exemplified above.

FINAL CONSIDERATIONS

In this chapter, a summary of the numerous biological activities of Reishi mushrooms that have been described and studied by different researchers around the world has been presented, emphasizing the use of *G. lucidum* and its derivatives for cancer treatment. In the Bioprocess and Agro-industry Plant at the University of Caldas, several research and development projects have been performed on *G. lucidum* mushrooms. These research efforts have been aimed at new cultivation alternatives (not only via solid-state culture but also by submerged fermentation) to increase the quality and productivity of the fungal fruiting bodies, and to develop extraction schemes for Reishi metabolites to evaluate their different biological activities. In this regard, a number of research issues and challenges have arisen, which are currently being addressed or will be studied by the authors of this book at the Universidad de Caldas. Some of consideration are listed, as follows:

- Development of new substrate formulations with lignocellulosic waste or easy-to-grow substitute materials for Reishi cultivation.
- Evaluation of the quality of *G. lucidum* culture in relation to the presence of contamination and competitors in the culture as well as assessments of the environmental conditions for mushroom growth and development, including the effects of different light wavelengths.
- Evaluation of the effects of the process conditions during the incubation phase on the biological efficiency of the culture.
- Evaluation of the quality of Reishi fruiting bodies obtained from different formulations containing alternative substrates.
- Influence of the feedstocks in the substrate mixtures on the fruiting of *G. lucidum.*
- Relationship between the nutritional content of Reishi carpophores and the nutritional content of the substrates in use.
- Establishment of *G. lucidum* cultivation techniques for tropical weather other than those that have already been developed.

- Study of the causes of contamination that directly affect Reishi culture and its relationship to the feedstocks and process conditions.
- Research and development on the use and transformation of spent substrates from Reishi cultivation into valuable products.
- Development of new medicinal products using Reishi fruiting bodies.
- Research and development on the extraction schemes of the biologically active substances from *G. lucidum* and on the techniques for determining the different biological activities.
- Research and development on the use of Reishi as a nutraceutical product (e.g., dietary supplements) in different groups of human populations to evaluate the influence of the constant intake of this mushroom.
- Evaluation of the recommended intake of Reishi nutraceuticals in different population groups that elicit a change in the test organisms.

REFERENCES

Aluko, R. E. 2012. *Functional Foods and Nutraceuticals*. Winnipeg, MB, Canadá: Springer.

Batbayar, S., D. H. Lee, and H. W. Kim. 2012. "Immunomodulation of fungal β-glucan in host defense signaling by dectin-1." *Biomolecules & Therapeutics* 20 (5):433-45.

Becerra-Jiménez, J. M., B. Valdivia-Urdiales, C. N. Aguilar, J. L. Martínez-Hernández, J. C. Contreras-Esquivel, and C. A. Cuellar-Lozano. 2011. "Quitosán fúngico: fuentes y propiedades." *Revista Científica de la Universidad Autónoma de Coahuila* 3 (5):24-32. ["Fungal chitosan: Sources and properties." *Scientific Journal of the Autonomous University of Coahuila* 3(5):24-32].

Belitz, H.-D., and W. Grosch. 1997. *Química de los Alimentos*. Zaragoza: Acribia. [*Food Chemistry*. Zaragoza: Acribia].

Bidegain, M., P. Postemsky, R. González Matute, D. Figlas, R. Devalis, S. Delmastro, C. Pereyra Huertas, N. Curvetto, and M. Cubitto. 2014. "*Optimización de la producción del hongo medicinal Reishi (Ganoderma lucidum) para el desarrollo de nutracéuticos y fitoterápicos.*" Universidad Nacional de Córdoba. ["*Optimization of the production of the medicinal mushroom Reishi (Ganoderma lucidum) for the development of nutraceuticals and phytotherapics.*" National University of Córdoba]. Accessed June 2017. https://rdu.unc.edu.ar/bitstream/handle/11086/2488/10659-28072-1-SM.pdf?sequence=1&isAllowed=y.

Bohn, J., and J. BeMiller. 1995. "(1→3)-β-D-Glucans as biological response modifiers: a review of structure-functional activity relationships." *Carbohydrate Polymers* 28 (1):3-14.

Byerrum, R., D. Clarke, E. Lucas, R. Ringler, J. Stevens, and C. Stock. 1957. "Tumor inhibitors in Boletus edulis and other Holobasidiomycetes." *Antibiotics & Chemotherapy (Northfield)* 7 (1):1-4.

Carlile, M., and S. Watkinson. 2000. *The Fungi*. London: Academic Press.

Cassileth, B. R. 2000. "Complementary therapies: the American experience." *Supportive Care in Cancer* 8 (1):16-23.

Chan, G. C.-F., W. K. Chan, and D. M.-Y. Sze. 2009. "The effects of β-glucan on human immune and cancer cells." *Journal of Hematology & Oncology* 2 (1):25.

Chang, S. T. 1991. "Cultivated mushrooms." In *Handbook of Applied Mycology. Foods and Feeds*, edited by D. K. Arora, K. G. Mukerji and E. H. Marth, 221-40. New York: Marcel Dekker.

Chang, S. T., and P. G. Miles. 2004. *Mushrooms. Cultivation, Nutritional Value, Medicinal Effect, and Enviromental Impact*. London: CRC Press.

Ćilerdžić, J., J. Vukojević, M. Stajić, T. Stanojković, and J. Glamočlija. 2014. "Biological activity of *Ganoderma lucidum* basidiocarps cultivated on alternative and commercial substrate." *Journal of Ethnopharmacology* 155 (1):312-19.

Collins, M. D., U. Rodrigues, C. Ash, M. Aguirre, J. A. E. Farrow, A. Martinez- Murcia, B. A. Phillips, A. M. Williams, and S. Wallbanks. 1991. "Phylogenetic analysis of the genus *Lactobacillus* and related lactic acid bacteria as determined by reverse transcriptase sequencing of 16S rRNA." *FEMS Microbiology Letters* 77 (1):5-12.

Curt, G. A. 1998. "Investment in research as a national priority." *Oncologist* 3 (1):64-66.

De Silva, D. D., S. Rapior, K. D. Hyde, and A. H. Bahkali. 2012. "Medicinal mushrooms in prevention and control of diabetes mellitus." *Fungal Diversity* 56 (1):1-29.

Dong, C., and Q. Han. 2015. "*Ganoderma lucidum* 灵芝 (Lingzhi, Ganoderma)." In *Dietary Chinese Herbs*, edited by Y. Liu, Z. Wang and J. Zhang, 759-66. Wien: Springer.

El Enshasy, H. A., and R. Hatti-Kaul. 2013. "Mushroom immunomodulators: unique molecules with unlimited applications." *Trends in Biotechnology* 31 (12):668-77.

Fukuzawa, M., R. Yamaguchi, I. Hide, Z. Chen, Y. Hirai, A. Sugimoto, T. Yasuhara, and Y. Nakata. 2008. "Possible involvement of long chain fatty acids in the spores of *Ganoderma lucidum* (Reishi Houshi) to its anti-tumor activity." *Biological and Pharmaceutical Bulletin* 31 (10):1933-37.

Gao, Y., E. Chan, and S. Zhou. 2004. "Immunomodulating activities of *Ganoderma*, a mushroom with medicinal properties." *Food Reviews International* 20 (2):123-61.

Gao, Y., S. Zhou, W. Jiang, M. Huang, and X. Dai. 2003. "Effects of Ganopoly® (A *Ganoderma lucidum* polysaccharide extract) on the immune functions in advanced- stage cancer patients." *Immunological Investigations* 32 (3):201-15.

Han, J. R., C. H. An, and J. M. Yuan. 2005. "Solid- state fermentation of cornmeal with the basidiomycete *Ganoderma lucidum* for degrading starch and upgrading nutritional value." *Journal of Applied Microbiology* 99 (4):910–15.

Hirahara, N., T. Edamatsu, A. Fujieda, M. Fujioka, T. Wada, and Y. Tajima. 2012. "Protein-bound polysaccharide-K (PSK) induces apoptosis via p38 mitogen-activated protein kinase pathway in promyelomonocytic leukemia HL-60 cells." *Anticancer Research* 32 (7):2631-37.

Hobbs, C. 1995. *Medicinal Mushrooms: An exploration of Tradition, Healing, and Culture*. Santa Cruz, CA, USA: Botanica Press.

Izydorczyk, M. 2005. "Understanding the chemistry of food carbohydrates." In *Food Carbohydrates: Chemistry, Physical Properties, and Applications*, edited by S. W. Cui, 1-68. Boca Raton, FL, USA: CRC Press.

Jiang, J., V. Slivova, K. A. Harvey, T. Valachovicova, and D. Sliva. 2004. "*Ganoderma lucidum* suppresses growth of breast cancer cells through the inhibition of Akt/NF-κB signaling." *Nutrition and Cancer* 49 (2):209-16.

Jin, X., J. Ruiz Beguerie, D. M. Y. Sze, and G. C. F. Chan. 2016. *Ganoderma lucidum* (Reishi mushroom) for cancer treatment. *Cochrane Database of Systematic Reviews* 4 (Art. No.: CD007731). Accessed June 2017. doi:10.1002/14651858.CD007731.pub3.

Kidd, P. 2000. "The use of mushroom glucans and proteoglycans in cancer treatment." *Alternative Medicine Review* 5 (1):4-27.

Kogan, G., and A. Kocher. 2007. "Role of yeast cell wall polysaccharides in pig nutrition and health protection." *Livestock Science* 109 (1-3):161-65.

Laurienzo, P. 2010. "Marine polysaccharides in pharmaceutical applications: an overview." *Marine Drugs* 8 (9):2435-65.

Lee, D. K., and C. Chang. 2003. "Expression and degradation of androgen receptor: mechanism and clinical implication." *The Journal of Clinical Endocrinology & Metabolism* 88 (9):4043-54.

Lehninger, A. L. 1981. *Bioquímica: las bases moleculares de la estructura y función celular*. Barcelona: Omega. [*Biochemistry: The Molecular Basis of Cellular Structure and Function*. Barcelona: Omega].

Lin, D., and Z. Lin. 2014. *Training Course on JUNCAO Technology for Latin America, Caribbean and South-Pacific Regions*. Edited by F. A. a. F. University, Fuzhou, China (Course documentation).

Liu, G.-Q., and K.-C. Zhang. 2007. "Enhancement of polysaccharides production in *Ganoderma lucidum* by the addition of ethyl acetate extracts from *Eupolyphaga sinensis* and *Catharsius molossus*." *Applied Microbiology and Biotechnology* 74 (3):572-77.

Lull, C., H. J. Wichers, and H. F. J. Savelkoul. 2005. "Antiinflammatory and immunomodulating properties of fungal metabolites." *Mediators of Inflammation* 2:63-80.

Mahajna, J., N. Dotan, B.-Z. Zaidman, R. D. Petrova, and S. P. Wasser. 2008. "Pharmacological values of medicinal mushrooms for prostate cancer therapy: The case of *Ganoderma Lucidum*." *Nutrition and Cancer* 61 (1):16-26.

Mantovani, M. S., M. F. Bellini, J. P. Angeli, R. J. Oliveira, A. F. Silva, and L. R. Ribeiro. 2008. "β-Glucans in promoting health: Prevention against mutation and cancer." *Mutation Research/Reviews in Mutation Research* 658 (3):154-61.

Meng, X., H. Liang, and L. Luo. 2016. "Antitumor polysaccharides from mushrooms: a review on the structural characteristics, antitumor mechanisms and immunomodulating activities." *Carbohydrate Research* 424:30-41.

Miura, T., L. Yuan, L. Yuan, B. Sun, and K.-i. Kosuna. 2003. "Isoflavone aglycon produced by culture of soybean extracts with basidiomycetes and its anti-angiogenic activity." *Bioscience, Biotechnology, and Biochemistry* 66 (12):2626-31.

Mizuno, T., G. Wang, J. Zhang, H. Kawagishi, and T. Nishitoba. 1995. "Reishi, *Ganoderma lucidum* and *Ganoderma tsugae*: Bioactive substances and medicinal effects." *Food Reviews International* 11 (1):151-66.

Montoya, S., D. M. López, and B. Segura. 2018. "Influencia de la luz azul sobre la productividad del cultivo sólido de *Ganoderma lucidum*." *Revista Colombiana de Biotecnología* 20 (1):51-58. ["Influence of blue light on the productivity of the solid culture of *Ganoderma lucidum*." *Colombian Journal of Biotechnology* 20 (1):51-58].

Montoya, S., Ó. J. Sánchez, and L. Levin. 2013. "Polysaccharide production by submerged and solid-state cultures from several medicinal higher basidiomycetes." *International Journal of Medicinal Mushrooms* 15 (1):71-79.

Moradali, M.-F., H. Mostafavi, S. Ghods, and G.-A. Hedjaroude. 2007. "Immunomodulating and anticancer agents in the realm of macromycetes fungi (macrofungi)." *International Immunopharmacology* 7 (6):701-24.

Müller, C. I., T. Kumagai, J. O'Kelly, N. P. Seeram, and D. Heber. 2006. "*Ganoderma lucidum* causes apoptosis in leukemia, lymphoma and multiple myeloma cells." *Leukemia Research* 30:841-48.

National Pharmacopoeia Committee. 2010. *The People's Republic of China Pharmacopoeia, Part 1*. Beijing: Chemical Industrial Publisher.

Ofodile, L. N., N. U. Kokubum, O. R. J. Grayer, O. T. Ogundipe, and M. S. J. Simmonds. 2005. "Antimicrobial activity of some *Ganoderma* species from Nigeria." *Phytotherapy Research* 19 (4):310-13.

Parkin, D. M., F. Bray, J. Ferlay, and P. Pisani. 2002. "Global cancer statistics." *CA: A Cancer Journal for Clinicians* 55 (2):74-108.

Paterson, R. R. M. 2006. "Ganoderma – A therapeutic fungal biofactory." *Phytochemistry* 67 (18):1985-2001.

Prajapati, V. D., G. K. Jani, and S. M. Khanda. 2013. "Pullulan: an exopolysaccharide and its various applications." *Carbohydrate Polymers* 95 (1):540-49.

Ramberg, J. E., E. D. Nelson, and R. A. Sinnott. 2010. Immunomodulatory dietary polysaccharides: a systematic review of the literature. *Nutrition Journal* 9 (54): 1-22. Accessed May 2018. doi:10.1186/1475-2891-9-54.

Rice, P. J., E. L. Adams, T. Ozment-Skelton, A. J. Gonzalez, M. P. Goldman, B. E. Lockhart, L. A. Barker, K. F. Breuel, W. K. DePonti, J. H. Kalbfleisch, H. E. Ensley, G. D. Brown, S. Gordon, and D. L. Williams. 2005. "Oral delivery and gastrointestinal absorption of soluble glucans stimulate increased resistance to infectious challenge." *Journal of Pharmacology and Experimental Therapeutics* 314 (3):1079-86.

Ross, G. D., V. Větvička, J. Yan, Y. Xia, and J. Větvičková. 1999. "Therapeutic intervention with complement and β-glucan in cancer." *Immunopharmacology* 42 (1-3):61-74.

Rossi, P., R. Difrancia, V. Quagliarello, E. Savino, P. Tralongo, C. L. Randazzo, and M. Berretta. 2018. "B-glucans from *Grifola frondosa* and

Ganoderma lucidum in breast cancer: an example of complementary and integrative medicine." *Oncotarget* 9 (37):24837-56.

Roy, S., M. A. A. Janan, K. K. Das, S. K. Munshi, and R. Noor. 2015. "Artificial cultivation of *Ganoderma lucidum* (Reishi Medicinal Mushroom) using different sawdusts as substrate." *American Journal of BioScience* 3 (5):178-82.

Sánchez, Ó. J., S. Montoya, and L. M. Vargas. 2015. "Polysaccharide production by submerged fermentation." In *Polysaccharides*, edited by K. G. Ramawat and J.-M. Mérillon, 451-73. Cham, Switzerland: Springer.

Smith, J. E., N. J. Rowan, and R. Sullivan. 2002. "*Medicinal Mushrooms: Their Therapeutic Properties and Current Medical Usage with Special Emphasis on Cancer Treatments.*" Glasgow: University of Strathclyde and Cancer Research UK.

Tomasi, S., F. D. Lohézic-Le, P. Sauleau, C. Bezivin, and J. Boustie. 2004. "Cytotoxic activity of methanol extracts from Basidiomycete mushrooms on murine cancer cell lines." *Pharmazie* 59 (4):290-93.

Vega, A., and H. Franco. 2013. "Productividad y calidad de los cuerpos fructíferos de los hongos comestibles *Pleurotus pulmonarius* RN2 y *P. djamor* RN81 y RN82 cultivados sobre sustratos lignocelulósicos." *Información Tecnológica* 24:69-78. ["Productivity and quality of the fruiting bodies of the edible fungi *Pleurotus pulmonarius* RN2 and *P. djamor* RN81 and RN82 grown on lignocellulosic substrates." *Technological Information* 24:69-78].

Wang, K.-p., J. Wang, Q. Li, Q.-l. Zhang, R.-x. You, Y. Cheng, L. Luo, and Y. Zhang. 2014. "Structural differences and conformational characterization of five bioactive polysaccharides from *Lentinus edodes*." *Food Research International* 62:223-32.

Wasser, S. P., and A. L. Weis. 1999. "Medicinal properties of substances ocurring in higher basidiomycetes mushromms: current perspectives (review)." *International Journal of medicinal Mushrooms* 1:31-62.

Wasser, S. P., I. V. Zmitrovich, M. Y. Didukh, W. A. Spirin, and V. F. Malysheva. 2006. *Morphological traits of Ganoderma lucidum complex highlighting G. tsugae var. jannieae: The current generalization.* Ruggell, Liechtenstein: ARA Gantner Verlag K.-G.

Wu, B., J. Cui, C. Zhang, and Z. Li. 2012. "A polysaccharide from *Agaricus blazei* inhibits proliferation and promotes apoptosis of osteosarcoma cells." *International Journal of Biological Macromolecules* 50:1116–20.

Yamanaka, D., R. Tada, Y. Adachi, K.-i. Ishibashi, M. Motoi, Y. Iwakura, and N. Ohno. 2012. "*Agaricus brasiliensis*-derived β-glucans exert immunoenhancing effects via a dectin-1-dependent pathway." *International Immunopharmacology* 14:311–19.

Yue, L., H. Cui, C. Li, Y. Lin, Y. Sun, Y. Niu, X. Wen, and J. Liu. 2012. "A polysaccharide from *Agaricus blazei* attenuates tumor cell adhesion via inhibiting E-selectin expression." *Carbohydrate Polymers* 88 (4):1326–33.

Zaidman, B.-Z., S. P. Wasser, E. Nevo, and J. Mahajna. 2008. "*Coprinus comatus* and *Ganoderma lucidum* interfere with androgen receptor function in LNCaP prostate cancer cells." *Molecular Biology Report* 35 (2):107-17.

Zaidman, B.-Z., M. Yassin, J. Mahajna, and S. P. Wasser. 2005. "Medicinal mushroom modulators of molecular targets as cancer therapeutics." *Applied Microbiology and Biotechnology* 67 (4):453-68.

Zhang, Y., H. Kong, Y. Fang, K. Nishinari, and G. O. Phillips. 2013. "Schizophyllan: A review on its structure, properties, bioactivities and recent developments." *Bioactive Carbohydrates and Dietary Fibre* 1 (1):53-71.

Zhu, H. S., X. L. Yang, L. B. Wang, D. X. Zhao, and L. Chen. 2000. "Effects of extracts from sporoderm-broken spores of *Ganoderma lucidum* on HeLa cells." *Cell Biology and Toxicology* 16 (3):201-06.

Zhu, X.-L., and Z.-B. Lin. 2005. "Effects of *Ganoderma lucidum* polysaccharides on proliferation and cytotoxicity of cytokine-induced killer cells." *Acta Pharmacologica Sinica* 26 (9):1130-37.

Ziliotto, L., F. Pinheiro, L. F. Barbisan, and M. A. Rodrigues. 2009. "Screening for in vitro and in vivo antitumor activities of the mushroom *Agaricus blazei*." *Nutrition and Cancer* 61 (2):245-50.

Zong, A., H. Cao, and F. Wang. 2012. "Anticancer polysaccharides from natural resources: A review of recent research." *Carbohydrate Polymers* 90 (4):1395-410.

Chapter 4

MARKET AND FEEDSTOCKS FOR MAITAKE CULTIVATION

INTRODUCTION

Grifola frondosa (synonym: *Polyporus frondosus* Dicks.: Fr) is known as Maitake ("dancing mushroom") in Japanese, Huī shù huā ("gray tree flower") in Chinese, and "hen of the woods" or "sheep's head mushroom" in North America (Chang and Miles 2004, Stamets 1993). This mushroom is a species in the class Agaricomycetes and belongs to the family Meripilaceae and the order Polyporales. The dancing mushroom name is explained by the fact that those who found this rare mushroom began dancing with joy; others attribute its name to the way the fruiting bodies of the mushroom overlap with one other, giving the appearance of dancing butterflies. *G. frondosa* is a large fleshy polypore; its fruiting body is 15 to 60 cm or more in size, and it consists of numerous small and overlapping shells arising from a common, fleshy and repeatedly branched base (Chang and Miles 2004). This mushroom is a white rot fungus, and it is in demand for its medicinal and edible attributes; it originates from deciduous forests in the temperate regions of northern North America, Europe and countries such as Japan and China. In addition, *G. frondosa* can be found in the subtropics, in high mountains with temperate climates (Chen et al. 2000). Three homologs of *G. frondosa* have been recorded. *Grifola albicans* is similar to *G. frondosa*, but it is white to pale yellow, fragile, and has a soft texture; in nature, it appears approximately one week before *G. frondosa* does. *Grifola gigantea* has petals that are comparatively larger than those of *G. frondosa*, measuring 10 to 20 cm in diameter; this species is edible and is used for medicinal purposes in China. *Grifola umbellatta* is an edible species, similar to *G. frondosa*, that has whitish gray to smoky brown petals and central stems; in China, this mushroom is used as a stimulant of the immune system, although it is not common in the wild (Chang and Miles 2004).

In this chapter, general aspects of the market for Maitake mushrooms are presented. In addition, some research cases describing the main conclusions of market studies performed under Colombian conditions are given. Finally, an overview on the specific raw materials employed for producing *G. frondosa* is provided.

General Aspects of the Market for Maitake Mushrooms

Global Market for Nutraceuticals

Humanity remains on a permanent search for new substances that can improve biological functions and, therefore, health in general. For several decades, Western society has turned to plants, herbs and food as sources of these health enhancers. These products have been named in various ways, including vitamins, dietary supplements, functional foods, nutraceuticals, phytochemicals (Zeisel 1999), and nutritional supplements (Chang and Buswell 1996). It is worth noting that these products have been designed to complement the human diet by increasing the intake of bioactive compounds to improve human health and physical conditions but not to be used as regular foods. The industries involved in providing these substances are expanding in the United States.

Notably, Eastern cultures have a long history of using traditional medicines associated with health foods in the forms of recognized nutritional foods, food supplements, medicinal herbs, and crude powdered drugs derived from plant, animal, mushroom and marine sources (Basu, Thomas, and Acharya 2007). India and China are the two most important countries known for producing traditional functional food products and nutraceuticals. Both of these nations have large populations, in particular in rural, remote and inaccessible areas, which are totally dependent on herbal remedies and other naturally available bioresources for treating common ailments and as general preventive and protective medications.

The term "nutraceutical" was coined from "nutrition" and "pharmaceutical" in 1989 by Stephen DeFelice, MD, founder and chairman of the Foundation for Innovation in Medicine (FIM) of the United States. According to DeFelice, a nutraceutical can be defined as "a food (or part of a food) that provides medical or health benefits, including the prevention and/or treatment of a disease" (Kalra 2003). Nutraceuticals include functional foods, nutritional supplements, sports drinks, and medically formulated foods (KPMG International Cooperative 2015). The term "functional food" refers to a food that is similar in appearance to or may be a conventional food that is consumed as part of a regular diet and is demonstrated to have physiological benefits and/or reduce the risk of chronic disease beyond its basic nutritional functions (Basu, Thomas, and Acharya 2007), e.g., vitamin-enriched food. However, a "dietary supplement" is a manufactured product that is intended to supplement the diet, and such a supplement contains one or more of the following dietary ingredients: a vitamin, a mineral, an herb or other botanical, an extract, an amino acid, a metabolite, a dietary substance, or a combination of these ingredients; a dietary supplement is intended for use by humans to supplement their diets by increasing their total daily intake of certain components. These products are intended for ingestion in the forms of pills, capsules, tablets, or liquids. Dietary supplements are not intended for use as conventional foods or as the sole item in a meal or diet. From this perspective, edible and medicinal mushrooms are considered as nutraceuticals and may be functional foods or make up parts of dietary supplements.

The motivation for purchasing or consuming nutraceutical products, nutritional supplements, functional foods, and others has traditionally focused on beliefs and knowledge about nutrition and attitudes such as the sociodemographic characteristics of consumers (Bogue, Coleman, and Sorenson 2005, Teratanavat, Salin, and Hooker 2005, Verbeke 2005). The consumers of this type of products are informed people who value technological advances and self-care.

The firm BCC Research (2018) estimated that the global nutraceutical market could reach 336.1 billion US dollars by 2023, from $230.9 billion in 2018, at a compound annual growth rate of 7.8% from 2018 to 2023. The share in sales of the nutraceutical industry by geographic region is as follows: United States 30%, Japan 22%, Europe 14%, and the rest of the world 34%. In the United States, nutraceuticals are mainly listed under the category of vitamins and dietary supplements. A range of regulations apply, such as the Dietary Supplement Health and Education Act (DSHEA) from 1994, which covers dietary supplements; the Federal Food Drug and Cosmetic Act, which covers all foods and food additives; and the Good Manufacturing Practice regulations from 2007. In Japan, foods for specified health uses require premarketing approval, and they include products containing dietary ingredients that have reported beneficial physiological effects and promote health. In the European Union, a complex range of regulations applies to nutraceuticals, depending on the product type; these products are primarily overseen by the European Food Safety Authority (EFSA), which also evaluates claims and has a public Register of Nutrition and Health Claims. China has a lengthy and costly registration process, with three main entities involved in policing the industry; the State Food and Drug Administration (SFDA) is in charge of dietary supplements and registration. In India, the required manufacturing oversight is less strong than it is for pharmaceuticals, which may impact Indian producers' ability to produce nutraceuticals for export (KPMG International Cooperative 2015).

One of the main concerns in the dietary supplement market is the quality and innocuity. For this reason, the development of appropriate standards for quality assurance for this type of products is required by both the industry and consumers. In the United States, the global public health organization NSF International and leaders from major retailers and manufacturers created the Global Retailer and Manufacturer Alliance (GRMA) in 2014 to develop consensus-based standards for dietary supplements. Combining retailer and regulatory requirements into a single standard and auditing program for each product category will help to reduce audits and costs while strengthening safety, quality and trust throughout the supply chain (NSF International 2014). The GRMA team is led by quality and safety experts from leading supermarkets, drugstore chains, mass merchandisers and club stores, including Ahold USA, Costco, Daymon Worldwide, Walgreens, Walmart, and Wegmans. This team developed a blueprint to create standardized requirements for each product category. Additionally, utilizing insights from industry experts such as the Council for Responsible Nutrition, American National Standards Institute (ANSI), Nutrition Business Journal and NSF International led to the creation of formal ANSI standard notifications, the first step in the standards development process. Undoubtedly, the development of these standards will greatly support the confidence of consumers of dietary supplements based on mushrooms, in turn stimulating the growth of the related market.

Medicinal mushrooms, such as *G. frondosa* and *G. lucidum*, are nutraceutical products. In particular, Wasser (2014) notes that medicinal mushrooms are used as functional foods, dietary supplements, a new class of drugs called "mushroom pharmaceuticals", natural biocontrol agents in plant protection, and "cosmeceuticals". Biocontrol agents based on mushrooms exhibit insecticidal, fungicidal, bactericidal, herbicidal, nematicidal, and antiphytoviral activities, while cosmeceuticals correspond to different compounds that are employed by the cosmetic industry for their film-forming capability, activation of epidermal growth factors, antioxidative, anti-allergic, antibacterial, and anti-inflammatory activities, stimulation of collagen activity, inhibition of autoimmune vitiligo, and treatment of acne. In particular, the market for dietary supplement products containing medicinal mushrooms or their parts is growing quickly and has a value of more than $18 billion per year.

Maitake Market

G. frondosa is an edible and medicinal mushroom known for its taste and health benefits. Maitake is a very large mushroom, and it grows deep in the mountains of northeastern Japan as well as in North America and Europe. The fruiting body of this mushroom is used in Japan, China, Korea, parts of Europe and North America as a regular food item or as a food additive due to its distinctive rich, woody flavor and meaty texture (Zhuang and Wasser 2004). *G. frondosa* fruiting bodies are large, soft-fleshed, and edible, and they consist of approximately 86-91% moisture and 9-14% dry matter, of which carbohydrates represent 59-60%, crude protein 21-22%, crude fiber 10%, crude fat 3–4%, and ash 5-7%; these fruiting bodies also contain ergosterol and assorted vitamins and minerals (Gregori et al. 2016). This mushroom is used in Chinese traditional medicine, in which it is called “keisho” and is used for treating hemorrhoids, improving spleen and stomach diseases, and curing various skin diseases. Currently, the production of processed foods such as tea and dietary supplements employ the dried fruiting body of *G. frondosa* as one of the ingredients. The biological properties of the Maitake fruiting body and liquid-cultured mycelia is of great interest in terms of marketing its functional foods (Dissanayake et al. 2018).

The Maitake has been grown commercially since the 1980s in Japan (Chang and Miles 2004), with significant market growth since then. Thus, by the end of the 1990s, approximately 40,000 tonnes were produced in Japan, and in 2001, nearly 14,600 tons of this fungus were produced in China. Because of this boom in Maitake consumption, in the early 1980s, a Japanese mycologist from Kobe Pharmaceutical University, Dr. Hiroaki Nanba, studied different medicinal mushrooms. He decided to focus on studying *G. frondosa*, and in 1984, he identified a fraction present in the mycelium and in the fruiting body that had antitumor activity when stimulating macrophages (Mayell 2001, Nanba, Hamaguchi, and Kuroda 1987). This fraction, which he named the D-fraction, consists of a mixture of proteins and polysaccharides, mainly β(1,6)-glucan and β(1,3)-glucan. In the 1990s, Nanba and Kubo continued to study the Maitake and managed to purify the D-fraction, which they called the MD-fraction, and which turns out to be more bioactive. In this way, they found that the D-fraction exhibits antitumor activity due to its potentiation of the immune response in the organism by polysaccharides (Mizuno et al. 2001, Nanba 1996).

Mushroom polysaccharides do not attack cancer cells directly but produce their antitumor effects by activating different immune responses in the host, usually as a result of the activation of macrophages due to their high molecular weight. Their mechanisms of action involve their recognition by several immune cell receptors as non-self molecules; thus, the immune system is stimulated by their presence. Structurally different β-glucans have different affinities towards receptors and thus generate different host responses (Chen and Seviour 2007). The immunomodulating and antitumor activities of these metabolites are related to immune cells, such as hematopoietic stem cells, lymphocytes, macrophages, T-cells, dendritic cells, and natural killer cells, which are involved in innate and adaptive immunity, resulting in the production of biologic response modifiers (Gregori et al. 2016, Moradali et al. 2007).

Maitake polysaccharides are stimulators of the immune system. Several researchers (Kodama, Komuta, and Nanba 2002, Kodama et al. 2002, Mizuno et al. 1986, Yamada, Nanba, and Kuroda 1990) experimentally found that there was a 75-85% regression of tumors in mice with carcinoma when the MD-fraction was administered intraperitoneally and orally. After this work, clinical trials were conducted in a hospital in Japan by administering Maitake MD-fraction tablets to patients after they stopped chemotherapy; a decrease in their tumor and tumor markers was obtained as well as an increase in the number of cells participating in the immune response (macrophages and lymphocytes). In February of 1998, the U.S. Food and Drug Administration approved an investigation using the D-fraction from Maitake mushrooms to treat advanced breast and prostate cancers (Mayell 2001). Different patents have been presented on the therapeutic use of *G. frondosa*, not only for preventing or treating cancer (e.g., Kubo and Nanba 2005, Tazawa, Odaira, and Watanabe 2003) but also for antidiabetic, antihypertensive, anti-obesity, and antihyperlipidemic effects (Zhuang, Kawagishi, and Harry 2007). Subsequently, a new polysaccharide from *G. frondosa*, called GFPS1b, was isolated, and it has antitumor activity in cancer cells in vitro (Cui, Li, et al. 2007, Cui, Tao, et al. 2007).

In general, Maitake mushrooms are used in different products such as dietary supplements for their biologically active properties, which allow for reductions in high levels of cholesterol and triglycerides in the blood (Kubo and Nanba 1997). The complex Maitake polysaccharides appear to act as immunomodulators, i.e., metabolites that increase the immune function. β-glucan, the primary immune-activating polysaccharide, is absorbed well when taken orally (Nanba 1993, Pengelly 1996). In addition, these immune-modulating polysaccharides (including β-glucan) may be supportive for people with HIV infections (Nanba 1992). People use Maitake mushrooms orally for HIV/AIDS, chronic fatigue syndrome (CFS), hepatitis, allergic rhinitis (hay fever), diabetes, hypertension, hyperlipidemia, weight loss, polycystic ovary syndrome (PCOS), immune system enhancement, cancer, myelodysplastic syndrome, and chemotherapy support (Natural Medicines Comprehensive Database 2018). The applicable parts of *G. frondosa* are the fruiting body and mycelium that contain β-glucan, which has been shown to possess antitumor activity. The D-fraction of β-glucan appears to be the most active and potent form. Maitake mushrooms have immunostimulant effects and activate natural killer cells, cytotoxic T-cells, interleukin-1, and superoxide anions. Maitake mushrooms can prevent the metastasis of experimentally induced tumors and can prevent tumor occurrence in normal cells. Researchers are particularly interested in Maitake mushrooms and their extracts because they appear to be orally active (Nanba 1993, Natural Medicines Comprehensive Database 2018). Other research suggests that these mushrooms can lower blood pressure and improve the lipid profile in experimental hyperlipidemia cases. Polysaccharides from the fruiting bodies of Maitake mushrooms also appear to have a hypoglycemic effect, possibly by activating insulin receptors. Since insulin resistance occurs in women with PCOS, there is also interest in using Maitake mushrooms for treating PCOS (Natural Medicines Comprehensive Database 2018).

Fresh and dried *G. frondosa* fruiting bodies are commonly available in the food markets of Asia, especially in Japan, China, and Korea. The fruiting bodies of this mushroom have also become available in other countries, such as the USA, Canada, and the EU (Gregori et al. 2016). In Japan, the Maitake is one of the most popular mushrooms along with the Shiitake (*Lentinus edodes*) and Enokitake (*Flammulina velutipes*), which are the best-selling mushrooms in wholesale markets. In Japan, mushrooms are an important ingredient in the diet and they also have a significant place in Japanese culture. The price of this mushroom falls in the upper middle group in the Japanese market, and it is sold at approximately 643 Yen/kg. Most Maitake mushrooms on the market are produced in Japan, and they are rarely imported from other countries. The Japanese Maitake market accounts for 320 million US dollars (Sun 2010). *G. frondosa* has quickly become a favorite mushroom in top restaurants across the country. Notably, numerous nutraceutical preparations based on powdered fruit bodies, dried extracts or purified polysaccharides from *G. frondosa* are emerging in different forms, such as capsules and tablets, and are being used as additives to food formulations (Gregori et al. 2016).

Maitake Mushroom Extract

The extract obtained from *G. frondosa* (mainly the D-fraction) is known as Maitake mushroom extract, which is commonly sold as a supplement. This extract is commercialized in different forms such as liquids, capsules, and powders. Maitake mushroom extract exhibits medicinal characteristics, including boosting the immune system, reducing blood pressure, chemotherapy control, etc., and it is used in foods such as salads, pizzas, soups, and fast food to enhance the health benefits of the food products. Nevertheless, Maitake mushroom extract may cause some side effects in people suffering from diabetes and low blood pressure and in pregnant women (Transparency Market Research 2018).

The market for Maitake mushroom extract is segmented based on its forms, nature (organic or with inorganic components), and application

(Transparency Market Research 2018). This extract can be commercialized in forms such as tablets, capsules, liquids, powders, etc. The most common forms in which this extract is sold is in powder and freeze-dried capsules. Considering the increased consumption of dietary supplements, there is increasing demand for Maitake extract because of its use in the global pharmaceutical industry in these powdered and liquid forms. Maitake mushroom extract may be applied by the food, cosmetic, and pharmaceutical industries. In the food industry, Maitake extract can be used as an ingredient in food products or it can be consumed along with a regular diet. In the cosmetic industry, this extract is used in the form of a cream or powder for treating skin diseases. In the pharmaceutical industry, Maitake mushroom extract is used to prepare medicines for treating health issues.

It is expected that the medicinal properties of the Maitake extract used for treating diseases such as high blood pressure, high cholesterol and cancer, as well as for supporting HIV treatment and chemotherapy, could drive the growth of its market in the pharmaceutical industry. Similarly, the use of Maitake extract for its health benefitting properties in different food products is driving the growth of the market in the food industry. However, the side effects of Maitake mushroom extract (it can lower blood pressure and blood sugar levels) may restrain market growth in the near future (Transparency Market Research 2018). The key players operating across the value chain of the global Maitake mushroom extract market include Nammex (North American Medicinal Mushroom Extracts), Solgar Inc., Mushroom Wisdom, Inc., Organika Health Products Inc., New Roots Herbal Inc., Na'vi Organics Ltd., Brain Forza Dietary Supplements LLC, Mushroom Science, etc.

Research Case 10. Market Research on G. frondosa Mushrooms (Global Context)

In 2011, the authors of the present book completed a research project entitled "Development and evaluation of the parameters of semi-industrial scale cultivation of the fungus *Grifola frondosa* in tropical climates and its transformation for commercial purposes as nutraceutical products" in Call

523-2011, "Project Bank for Creation of Technology-based Companies or Business Units", which was funded by the Colombian Administrative Department of Science, Technology and Innovation (Colciencias). This project was made possible by considering the results obtained in a previous R&D project entitled "Selection, Development and Stabilization of an Inoculum of the Macromycete *Grifola frondosa*" that was developed between 2006 and 2008 at the Universidad de Caldas. To achieve the goal of scaling-up the solid-state cultivation process for the industrial production of *G. frondosa* mushrooms in tropical climates, it was necessary to generate knowledge about the market for this mushroom in different forms and presentations.

From the market research for the project performed by specialized firms, a report from the International Society for Mushroom Science indicated that the global consumption of thirty different mushroom species was 3 million tonnes in 2002; of this, 2 million tonnes were cultivated, and 1 million tonnes were wild. The most representative presentations of the marketed mushrooms were as follows: dried, canned, and frozen mushrooms in plastic packaging and tin cans. There was an ascending line of production and consumption of mushrooms observed worldwide. In 2007, more than 7 million tonnes of fresh mushroom were produced with a value exceeding 30 billion US dollars.

This research indicated that Asia is the highest producer of fungi and mushrooms in the world, with a 49% share, followed by Europe (34%) and the Americas (16%). Between 2005 and 2010, Asia exhibited an annual production growth rate of 15%. Additionally, the primary importer of mushrooms between 2001 and 2010 was Germany with 16%, followed by Japan (9%) and the United States (6%).

Regarding Maitake production, this mushroom has spread to the United States, with the companies Mushroom Wisdom, Inc. (formerly Maitake Products, Inc.), Fungi Perfecti, and Mushroom Science being the most important players in the market. These companies commercialize various medicinal products from Asia and Europe such as Maitake Gold 404®, which is publicized as a potent liquid *G. frondosa* extract developed by Dr.

Hiroaki Nanba that is the only Maitake supplement to receive Dr. Nanba's personal endorsement.

Currently, *G. frondosa* is available for sale in different forms, ranging from the fruiting body or dry powdered mycelium and encapsulated in tea or more elaborate products, such as some of its active ingredients in isolated form. The active ingredients with greater distribution are the Maitake D-fraction (a polysaccharide), ethanolic extracts (primarily isolates of carbohydrates and terpenoids), and other non-alcoholic liquid extracts that generally correspond to aqueous extracts of the dry fruiting bodies. Maitake mushrooms are available through the websites of companies that produce medicinal mushrooms and those of marketers of natural products, nutraceuticals, functional foods, and others, as well as more specialized web pages related to medical and agricultural sciences. However, the Maitake market is still quite constrained globally due to limited production on a commercial scale. The highest commercial cultivation of Maitake mushrooms in the world is performed in countries such as Japan, the United States, China, and some European countries. These producers do not meet the global demand for the mushroom and its derivatives, which is increasing due to its range of medicinal attributes.

Maitake Market in Colombia

In Colombia, the segment of nutraceutical products, dietary supplements, and functional foods is growing steadily based on the part of the population that seeks alternative and natural ways to improve their health and living conditions as well as to apply complementary therapies for treating different diseases, among other reasons. Dry mushrooms and derived products, nutraceutical products, natural extracts, etc. are classified as herbal products and natural products for homeotherapy, alternative medicine, health food stores, and complementary medical treatments, among others. According to Procolombia, the governmental agency in charge of promoting tourism and foreign investment in Colombia, for non-mining energy exports and the image of the country, the importation of natural products, among which are dry mushrooms, functional fungal foods, nutritional supplements, and others, is not yet consolidated or sufficiently established as an important item in the Colombian market. It is noteworthy that mushrooms and fungi that originated in Colombia are admitted into the United States through all access ports, as long as they comply with the requirements for fresh and processed products and the regulations established by the United States Department of Agriculture.

The Maitake market in Colombia is limited to the quantity of the mushroom or its derived products that the country can import since there are no established commercial Maitake solid-state cultivation facilities, nor mycelial production by other production methods. In Colombia, there is no direct competition for products based on Maitake, only substitute products. It is clear that the differentiating value of this type of fungal products relies on its scientific basis. In a preliminary analysis, it was found that there is a supply of similar products abroad and that the differentiation is based on two aspects, the scientific basis and the quality of the product.

Regarding products based on *G. frondosa* within the country, there is no detailed information available except for the studies and research that have been conducted at the Universidad de Caldas. However, different products that could become substitutes (especially imported products) of Maitake-based products are being offered on the market. From the surveys conducted to date, when referring to the distribution channels of the nutraceutical industry, it has been determined that these products and substitutes are commercialized through specialized stores, drug stores or multilevel sales in the same way the industry does for recognized brands.

In the following research case, the market aspects related to *G. frondosa* and products based on this mushroom for Colombia are briefly explained by considering the outcomes of the project "Development and evaluation of the parameters of semi-industrial scale cultivation of the fungus *Grifola frondosa* in tropical climates and its transformation for commercial purposes as nutraceutical products".

Research Case 11. Aspects of the Maitake Market in Colombia

Colombia presents exceptional geographic and climatic settings, and it has emerged as a country with privileged conditions for producing a large number of natural agricultural, agro-industrial, and industrial products through biotechnology. The commercial solid-state cultivation of mushrooms with medicinal attributes such as Maitake has already been tested and standardized by researchers from the Universidad de Caldas in Manizales, Colombia. This is how biological efficiencies of Maitake production between 30% and 35% (see equation 1) were attained, which are competitive in the market. From the different research and development efforts conducted for cultivating *G. frondosa* in tropical climates such as that of Manizales, a city located in the mountainous Andean region of Colombia, the complete scheme for producing this mushroom at the industrial scale has already been developed, including the definition of the requirements for the equipment, machinery, processing areas, feedstock supply, etc. These basic and detailed engineering developments are significant because they are necessary to exploit Colombia's potential as a promising country for producing the Maitake mushroom in Latin America on a commercial scale to meet not only the domestic demand but also the demand from other countries in the region.

From the market research, Colombia was found to import fresh and refrigerated mushrooms under tariff heading HS0709510000. However, this tariff heading includes all fungal species, so it is not possible to discriminate among them by species. From the information obtained from the Ministry of Industry, Tourism and Commerce, the countries to which Colombia exports fresh and refrigerated mushrooms are Guatemala, Panama, Costa Rica, the Netherlands Antilles, Mexico, the United States, Nicaragua, Ecuador, France, and Chile. The main export companies are Setas de Colombia S.A., Champiñones Potín, and Colombiana de Exportaciones Agropecuarias S.A. These firms mainly export button mushrooms (*Agaricus bisporus*) in small quantities ranging from 5 kg to 25 tonnes per month. Regarding the *G. frondosa* species, there are no data that provide information either about production or about the market. However, Colombia imported mushrooms, especially from China: 296 and 260 tonnes in 2010 and 2011, respectively. A similar behavior was noted for button mushroom imports from Chile and the European Union, probably because exports of this mushroom from Colombia to other countries increased from 2010 to 2011. For example,

Colombia exported 362 tonnes of fresh button mushrooms to the United States in 2010 and 507 tonnes in 2011.

During the market survey, written questions were submitted via e-mail to global distributors of *A. bisporus* and other mushrooms to gauge their level of knowledge and their commercialization of *G. frondosa* mushrooms. The collected information indicated that the Maitake is a fungus with great commercial limitations, which ratifies the limitations in its production on an industrial scale. Therefore, this mushroom is presented as promising for a country such as Colombia. In addition, the market research included a sample to investigate the consumption of Maitake mushrooms in the market segments of pharmaceutical products, nutraceutical foods, and dietary supplements. Thus, for the pharmaceutical and edible categories, the respondents answered that they know the attributes and benefits of several macromycete species, but they have not worked with nor do they know the fungus *G. frondosa*. Similarly, those who work for the dietary supplement market largely affirmed their lack of knowledge of the Maitake mushroom; however, those who do know it do not commercialize this mushroom because they have not found marketing channels or stable suppliers of the Maitake or its derived products.

To complete the market research, a proof of concept study should be performed. This proof should consist of examining or evaluating the attributes of a product. Therefore, these tests are performed to clarify the behavior or reaction of consumers of a new product. In addition, these trials are intended to demonstrate the authenticity or lack thereof of the product. The objective of a proof of concept study is necessary to assess the concept of the product before starting mass production. The concept is the product idea developed in detail. In this way, a proof of concept corresponds to the search for the relationship between the new product and how it reaches the consumer, so its aim is to establish whether the consumer understands the concept of the new product. In the following research case, the development of the proof of concept for a Maitake product in Colombia will be explained.

Research Case 12. Proof of Concept for a Minimally Processed Maitake-Based Product

As a complement to the market research, a proof of concept was performed for a *G. frondosa* product in three cities in Colombia. The aim of this proof was to collect and analyze information on the appreciation, observations and recommendations of the target public about the presentation of a product based on dried, shredded and encapsulated *G. frondosa* mushrooms to make improvements on such a product and estimate its acceptance. The proof was performed in Manizales, Pereira, and Armenia, which are middle-sized Colombian cities (with populations between 250,000 and 600,000 inhabitants) located in the region of influence near the Universidad de Caldas. The target public to which the proof of concept was applied was divided into two groups. The first group included representatives, owners or commercial directors of health food shops as well as physicians engaged with alternative medicine, whereas the second group included the customers of health food stores. For this purpose, the concept of the product was defined as a dietary supplement containing 100% of dry Maitake mushrooms that contributes to the strengthening of the immune system.

For this proof of concept, an interview consisting of 14 questions focused on understanding the general aspects of interest for the new product, as follows:

- Interest in this product compared to others
- Interest in the product as presented
- Characteristics that are sought or identified to understand that the product is 100% natural
- Opinion on the packaging
- Opinion on the color of the packaging
- Opinion on the product image
- Opinion on the name of the product
- Suggestions for changes in the presentation of the product

- Aspects identified as positive for the product
- Aspects to be considered to make changes in the product
- Information on the purchase of natural products and the types of such products
- Purchase intention associated with the product
- Does the costumer consider that the product offers big advantages for his/her health?
- Does the costumer consider that the product is novel?

After conducting 260 interviews, the most relevant factor among the interviewees was found to be the function of the product (enhancer of the immune system); that is, they were more interested in the composition of the product than in the other characteristics. This type of interview can also be applied to determine the real purchase intentions of consumers of new natural products, as in the case of Maitake-based products in Colombia.

FEEDSTOCKS FOR PRODUCING *G. FRONDOSA*

Several researchers from around the world have performed evaluations of different agricultural and agro-industrial feedstocks and waste for the efficient production of Maitake mushrooms. Song et al. (2018) evaluated different mixtures of materials from various agro-industrial residues such as oak sawdust, corn cobs, corn straw, soy straw, and rice straw supplemented with wheat bran and calcium carbonate; all the mixtures were brought up to 62.5% moisture.

These authors prepared 21 mixtures with C/N ratios from 14.9 to 74.7 and found that formulations with C/N ratios between 25 and 29 exhibited the highest biological efficiency (*BE*), reaching values ranging from 32% up to 42%, while formulations with higher C/N ratios (between 55 and 74.7) yielded the lowest *BE* (below 30%).

Like these researchers, many others have shown that the raw materials in which *G. frondosa* develops properly to yield fruiting bodies are lignocellulosic materials, especially non-aromatic hardwoods such as oak trees, and they are supplemented with cereal brans from corn, wheat, rice, and others, calcium carbonate, and calcium sulfate with the use of initiators such as sucrose (Chang and Miles 2004, Montoya 2008, Montoya, Varón, and Levin 2008, Shen and Royse 2001, 2002, Stamets 2000). In particular, Stamets (1993) proposed utilizing larch sawdust supplemented with rice bran as a culture medium to produce Maitake fruiting bodies. In addition, this author recommends similar formulations to those used for the solid-state cultivation of *Lentinus edodes*, i.e., making combinations of different types of non-aromatic hardwoods with cereal bran and supplementing with calcium salts.

Chapter 1 provides a detailed description of the composition of lignocellulosic materials required for the growth and development of different cultivable macromycetes, among which are *G. frondosa* mushrooms. That chapter also provides a list of the main lignocellulosic waste available in Colombia that have been used for producing *Ganoderma lucidum* mushrooms and that could be employed for Maitake production as well.

It should be noted that the preferred lignocellulosic materials suitable for the solid-state cultivation of *G. frondosa* are oak sawdust and shavings (see Figure 29), corn bran (see Figure 30) or the bran of other cereals (rice or wheat), and chopped corn cobs, which are supplemented with calcium carbonate, calcium sulfate, and sucrose.

(a) (b)

Figure 29. Oak waste. (a) Oak sawdust; (b) oak chips (shavings).

Figure 30. Corn bran.

In the literature, different formulations of substrates for solid-state Maitake cultivation are presented while considering the availability of lignocellulosic materials in each country. For example, in Australia, adaptations of the cultivation have been made by using eucalyptus sawdust for the establishment of Maitake cultivation on a commercial scale (Chen et al. 2000, Sttot and Mohammed 2004). In the following research case, recommended formulations are provided while accounting for the raw materials available in tropical countries such as Colombia.

Research Case 13. Formulation of Solid Substrates for Maitake Cultivation

From the results of numerous cultivation tests performed at different scales at the Bioprocess and Agroindustry Plant at Universidad de Caldas, three variants of solid culture media were defined for an efficient production of *G. frondosa* mushrooms. The formulations of these culture media are presented in Table 25. Oak waste and chopped corn cobs are employed as carbon sources; the nitrogen source is cereal bran, and sucrose acts as an initiator for mycelium running.

After finding the mass balance on a dry basis, i.e., after calculating the amount of each component required for the preparation of the substrate, the water volume to add is calculated by accounting for the moisture content of each feedstock included in the formulation; it must be adjusted so that the solid medium contains up to 58% -60% moisture before being packed in the bags.

Table 25. Recommended substrate formulation for the solid-state cultivation of *Grifola frondosa* mushrooms obtained at the Bioprocess and Agroindustry Plant at the Universidad de Caldas

Component	Concentration (% dry basis)		
	I	II	II
Oak sawdust	74-75	55	55
Oak shavings		20	
Chopped corn cobs (0.5 cm in diameter)			20
Cereal (corn, rice or sorghum) bran	23	23	23
$CaCO_3$	1	1	1
$CaSO_4$	1-0		
Sucrose	1	1	1

FINAL CONSIDERATIONS

The global mushroom trade exhibits a growing trend, especially because many segments of the world's population tend to consume natural and health-benefitting products both as part of their diets and as supplementary products, such as nutritional supplements, active components, and others.

The consumption of *G. frondosa* mushrooms and derived products, such as Maitake mushroom extract, is important in some countries in Asia (primarily Japan, China, and Korea), Europe, and North America. However, processed or minimally processed Maitake products are also available in Latin America through the websites of companies that specialize in the commercialization of mushrooms with medicinal attributes. In Colombia, *G. frondosa* is a little-known and exotic mushroom. The cultivation technology for this fungus under Colombian conditions has been developed in the Bioprocess and Agro-industry Plant at the Universidad de Caldas. The knowledge generated there is offered to the industrial sector and rural organized communities for technology transfer purposes and to perform innovative activities that encourage the production of Maitake mushrooms, for which a significant potential market has been determined.

The raw materials available for cultivating *G. frondosa* mushrooms are mostly residues of non-aromatic hardwoods. In Colombia, the most highly available material is oak waste (sawdust or shavings). Various studies have shown that coffee residues should not be used because their high zinc content inhibits the fruiting of this mushroom. This issue is explained comprehensively in Chapter 6.

REFERENCES

Basu, S. K., J. E. Thomas, and S. N. Acharya. 2007. "Prospects for growth in global nutraceutical and functional food markets: A Canadian perspective." *Australian Journal of Basic and Applied Sciences* 1 (4):637-649.

BCC Research. 2018. "*Nutraceuticals: Global Markets to 2023.*" BCC Research. Accessed December 2018. https://www.bccresearch.com/market-research/food-and-beverage/nutraceuticals-global-markets-to-2023-report-fod013g.html.

Bogue, J., T. Coleman, and D. Sorenson. 2005. "Determinants of consumers' dietary behaviour for health-enhancing foods." *British Food Journal* 107 (1):4-16.

Chang, S. T., and J. A. Buswell. 1996. "Mushroom nutriceuticals." *World Journal of Microbiology and Biotechnology* 12:473–476.

Chang, S. T., and P. G. Miles. 2004. *Mushrooms. Cultivation, Nutritional Value, Medicinal Effect, and Enviromental Impact*. London: CRC Press.

Chen, A. W., P. Stamets, R. B. Cooper, N. Huang, and S. Han. 2000. "Ecology, morphology, and morphogenesis in nature of edible and medicinal mushroom *Grifola frondosa* (Dicks.: Fr.) S.F.Gray—maitake (Aphyllophoromycetideae)." *International Journal of Medicinal Mushrooms* 2 (3):221-228.

Chen, J., and R. Seviour. 2007. "Medicinal importance of fungal β-(1→ 3),(1→ 6)-glucans." *Mycological Research* 111 (6):635-652.

Cui, F. J., Y. Li, Y. Y. Xu, Z. Q. Liu, D. M. Huang, Z. C. Zhang, and W. Y. Tao. 2007. "Induction of apoptosis in SGC-7901 cells by polysaccharide-peptide GFPS1b from the cultured mycelia of *Grifola frondosa* GF9801." *Toxicology in Vitro* 21:417-427.

Cui, F. J., W. Y. Tao, Z. Xu, W. J. Guo, H. Y. Xu, Z. H. Ao, J. Jin, and Y. Q. Wei. 2007. "Structural analysis of anti-tumor heteropolysaccharide GFPS1b from the cultured mycelia of *Grifola frondosa* GF9801." *Bioresource Technology* 98:395-401.

Dissanayake, A. A., C.-R. Zhang, G. L. Mills, and M. G. Nair. 2018. "Cultivated maitake mushroom demonstrated functional food quality as determined by in vitro bioassays." *Journal of Functional Foods* 44:79-85.

Gregori, A., M. Švagelj, D. Voglar, and M. Berovic. 2016. "Growth characteristics and ergosterol content of *Grifola frondosa* in various solid-state substrates." *Chemical and Biochemical Engineering Quarterly* 30 (2):183-188.

Kalra, E. K. 2003. "Nutraceutical - Definition and Introduction." *AAPS PharmSci* 5 (3):27-28.

Kodama, N., K. Komuta, and H. Nanba. 2002. "Can maitake MD-fraction aid cancer patiens?" *Alternative Medicine Review* 7:236-239.

Kodama, N., K. Komuta, N. Sakai, and H. Nanba. 2002. "Effects of D-fraction, a polysaccharide from *Grifola frondosa* on tumor growth involve activation of NK cells." *Biological and Pharmaceutical Bulletin* 25:1647-1650.

KPMG International Cooperative. 2015. "*Nutraceuticals: The future of intelligent food. Where food and pharmaceuticals converge.*" KPMG International Cooperative. Accessed December 2018. https://home.kpmg.com/content/dam/kpmg/pdf/2015/05/neutraceuticals-the-future-of-intelligent-food.pdf.

Kubo, K., and H. Nanba. 1997. "Anti-hyperliposis effect of maitake fruit body (*Grifola frondosa*). I." *Biological and Pharmaceutical Bulletin* 20 (7):781-785.

Kubo, K., and H. Nanba. 2005. "Antitumor substance extracted from *Grifola frondosa*." Germany patent 69730420.

Mayell, M. 2001. "Maitake extracts and their therapeutic potencial." *Alternative Medicine Review* 6:48-60.

Mizuno, M., K. I. Minato, S. Kawakami, S. Tatsuoka, Y. Denpo, and H. Tsuchida. 2001. "Contents of anti-tumor polysaccharides in certain mushrooms and their immunomodulating activities." *Food Science and Technology Research* 7:31-34.

Mizuno, T., K. Ohsawa, N. Hagiwara, and R. Kuboyama. 1986. "Fractionation and characterization of antitumor polysaccharides from maitake, *Grifola frondosa*." *Agricultural and Biological Chemistry* 50:1679-1688.

Montoya, S. 2008. "*Actividad enzimática, degradación de residuos sólidos orgánicos y generación de biomasa útil del macromiceto Grifola frondosa*." MSc thesis, Departamento de Ingeniería Química, Universidad Nacional de Colombia sede Manizales. ["*Enzymatic activity, degradation of solid organic residues, and generation of useful biomass of the macromycete Grifola frondosa*." MSc thesis, Department of Chemical Engineering, National University of Colombia at Manizales].

Montoya, S., L. M. Varón, and L. Levin. 2008. "Effect of culture parameter on the production of the edible mushroom *Grifola frondosa* (maitake) in tropical weathers." *World Journal of Microbiology and Biotechnology* 24:1361-1366.

Moradali, M.-F., H. Mostafavi, S. Ghods, and G.-A. Hedjaroude. 2007. "Immunomodulating and anticancer agents in the realm of macromycetes fungi (macrofungi)." *International Immunopharmacology* 7 (6):701-724.

Nanba, H. 1992. "Immunostimulant activity in vivo and anti-HIV activity in vitro of 3 branched β-1-6-glucans extracted from maitake mushrooms (*Grifola frondosa*)." VIII International Conference on AIDS, Amsterdam, The Netherlands.

Nanba, H. 1993. " Antitumor activity of orally administered D-fraction from Maitake mushroom (*Grifola frondosa*)." *Journal of Naturopathic Medicine* 4:10-15.

Nanba, H. 1996. "El hongo Maitake, potente anticancerígeno y adaptógeno." *Natura Medicatrix* 43:1-4. ["The Maitake mushroom, potent anticancer and adaptogen." *Natura Medicatrix* 43:1-4].

Nanba, H., A. Hamaguchi, and H. Kuroda. 1987. "The chemical structure of an antitumor polysaccharide in fruit bodies of *Grifola frondosa* (Maitake)." *Chemical and Pharmaceutical Bulletin* 35:1162–1168.

Natural Medicines Comprehensive Database. 2018. "*Maitake Mushroom.*" Therapeutic Research Faculty. Accessed November 2018. http://naturaldatabase.therapeuticresearch.com/(X(1)S(zj4gik55kgf0a1z4ww100wnv))/nd/PrintVersion.aspx?id=560&AspxAutoDetectCookieSupport=1.

NSF International. 2014. "*NSF Partners with Global Retailer and Manufacturer Alliance to Develop Consumer Product Standards.*" NSF International. Accessed December 2018. https://www.nsf.org/en/newsroom/nsf-partners-with-global-retailer-and-manufacturer-alliance-to-develop-cons.

Pengelly, A. 1996. "Medicinal fungi of the world." *Modern Phytotherapist* 2:1-3.

Shen, Q., and D. J. Royse. 2001. "Effect of nutrient supplements of biological efficiency, quality of crop cycle time of maitake (*Grifola frondosa*)." *Applied Microbiology and Biotechnology* 57:74-78.

Shen, Q., and D. J. Royse. 2002. "Crop cycle time, yield and quality of maitake *Grifola frondosa* as influenced by nutrient supplements." *4th International Conference for Mushroom Biology and Mushroom Products* (The 4th ICMBMP), Cuernavaca, Mexico.

Song, B., J. Ye, F. L. Sossah, C. Li, D. Li, L. Meng, S. Xu, Y. Fu, and Y. Li. 2018. Assessing the effects of different agro-residue as substrates on growth cycle and yield of *Grifola frondosa* and statitical optimization of substrate components using simplex-lattice design. *AMB Express* 8 (46). Accessed November 2018. doi:10.1186/s13568-018-0565-8.

Stamets, P. 1993. *Growing Gourmet and Medicinal Mushrooms*. Hong Kong: Ten Speed Press and Mycomedia.

Stamets, P. 2000. *Growing Gourmet and Medicinal Mushrooms*. Berkeley: Ten Speed Press.

Sttot, K., and K. Mohammed. 2004. "*Specialty Mushroom Production System: Maitake and Morels*." Report 04/024. Barton ACT, Australia: Rural Industries Research and Development Corporation.

Sun, H. 2010. "*Competitive strategies for Chinese mushroom export to the Japanese market*." MSc thesis, Department of Economics, Swedish University of Agricultural Sciences.

Tazawa, K., Y. Odaira, and M. Watanabe. 2003. "*Active oxygen scavenger and cancer chemopreventer from Grifola*." United States patent 6616928.

Teratanavat, R., V. Salin, and N. H. Hooker. 2005. "Recall event timing: Measures of managerial performance in U.S. meat and poultry plants." *Agribusiness: an International Journal* 21 (3):351-373.

Transparency Market Research. 2018. "*Maitake Mushroom Extract Market - Global Industry Analysis, Size, Share, Growth, Trends and Forecast 2017 - 2025*." Transparency Market Research. Accessed December 2018. https://www.transparencymarketresearch.com/maitake-mushroom-extract-market.html.

Verbeke, W. 2005. "Consumer acceptance of functional foods: Socio-demographic, cognitive and attitudinal determinants." *Food Quality and Preference* 16 (1):45-57.

Wasser, S. P. 2014. "Medicinal mushroom science: Current perspectives, advances, evidences, and challenges." *Biomedical Journal* 37 (6):345-356.

Yamada, Y., H. Nanba, and H. Kuroda. 1990. "Antitumor effect of orally administered extracts from fruit body of *Grifola frondosa* (maitake)." *Chemotherapy* 38:790-796.

Zeisel, S. H. 1999. "Regulation of "Nutriceuticals"." *Science* 285:1853-1855.

Zhuang, C., C. Kawagishi, and G. Harry. 2007. "*Glycoprotein with antidiabetic, antihypertensive, antiobesity and antihyperlipidemic effects from Grifola frondosa and a method for preparing same*." United States patent 7214778.

Zhuang, C., and S. P. Wasser. 2004. "Medicinal Value of Culinary-Medicinal Maitake Mushroom *Grifola frondosa* (Dicks.:Fr.) S.F. Gray (Aphyllophoromycetideae). Review." *International Journal of Medicinal Mushrooms* 6 (4):28.

Chapter 5

Solid-State Cultivation of *Grifola frondosa* at Small and Pilot Scales

Introduction

Grifola frondosa (Maitake) has been artificially and commercially grown since the mid-1980s in Japan. Since then, the demand for this fungus has increased due to its important medicinal attributes. However, Maitake cultivation is not simple because it requires strict control over environmental conditions in addition to restrictions regarding the raw materials used for its cultivation. These restrictions in the culture media for Maitake growth and development are probably related to the fact that, in the wild, this mushroom grows on the base of dead trees or on trees during their dying process. The species of wood on which *G. frondosa* grows naturally are oak, elm, maple, black gums, beech, and chestnut trees. In the forests of North America and China, the fruiting bodies of wild mushrooms have been found to weigh up to 37 kg. Fungal species such as *G. frondosa* are the most easily recognized in the wild, as their physical characteristics are quite specific. For example, this mushroom exhibits multiple layers, heads or petals superimposed on a spoon shape that detaches from a single stem, which is reminiscent of a hairy hen (Chen et al. 2000).

The materials used for the artificial solid-state cultivation of *G. frondosa* are specifically based on waste from the wood species mentioned above. However, Shen and Royse (2001) tested materials other than wood waste for Maitake cultivation, such as husks from cotton, corn, rice, and sorghum, to replace part of the wood sawdust used in the substrates for artificial *G. frondosa* cultivation.

BIOLOGY AND FEATURES OF *GRIFOLA FRONDOSA*

Biology and Taxonomy of *G. frondosa*

Grifola frondosa (Dicks.) Gray, or hen of the woods, belongs to the phylum Basidiomycota, class Agaricomycetes, order Polyporales, family Meripilaceae, genus *Grifola*. Scottish mycologist James J. Dickson (1738-1822) described this polypore in 1785, establishing its basionym when he gave it the scientific name *Boletus frondosus*. It was another British scientist, Samuel Frederick Gray (1766-1828), who transferred this species to the genus *Grifola* in 1821, establishing its currently accepted scientific name as *Grifola frondosa*. Over the last two centuries, mycology researchers have described and given different scientific names to this fungus, with many synonyms such as *Boletus frondosus* Dicks., *Boletus elegans* Bolton, *Polyporus frondosus* (Dicks.) Fr., *Polyporus intybaceus* Fr., *Grifola frondosa* f. intybacea (Fr.) Pilát, and *Grifola intybacea* (Fr.) Imazeki.

Commonly, the fruiting body of *G. frondosa* is relatively short-lived compared to most polypores and is usually found at the base of old or dead oaks. Its appearance on other hardwood trees is less frequent. This mushroom generates a broad connection with the roots of the tree (host) where it develops. It is notable that *Grifola* spp. attack living trees, causing white rotting, and it smells unpleasantly when it decomposes. This macromycete is uncommon and has been found in Great Britain and Ireland, continental Europe and North America. Additionally, over the last century, *G. frondosa* has been found in the forests of Japan and China (O'Reilly 2016).

Figure 31. Fruiting bodies of *Grifola frondosa* grown on artificial solid culture and fructified in the wild.

Morphological Features of *G. frondosa*

The fruiting bodies of *G. frondosa* are sweet-smelling when young; this smell diminishes with age. The older specimens can be quite pungent. The carpophores consist of circular levels of a common branched stem. This polypore is composed of a kind of petal (frond) in the form of soft tabs that form a rosette in the shape of cauliflower and measures 20 to 50 cm wide. The individual petals or fronds measure 4 to 10 cm wide and 5 to 10 mm thick, and they vary in color from cinnamon to olive, gray, or cream in concentric undulating areas (see Figure 31). Very occasionally, *G. frondosa* presents in an almost pure white shape. The Maitake has white tubes with a depth of 2 to 3 mm that are generally rounded and end in pale cream pores, which are decurrent to the stem (see Figure 32). Maitake spores are widely ellipsoidal, inamyloid (not turning red or blue when stained with Melzer's reagent), smooth, and they measure 5.0-7.0 × 3.5-5.0 μm (Ainsworth 2008, O'Reilly 2016).

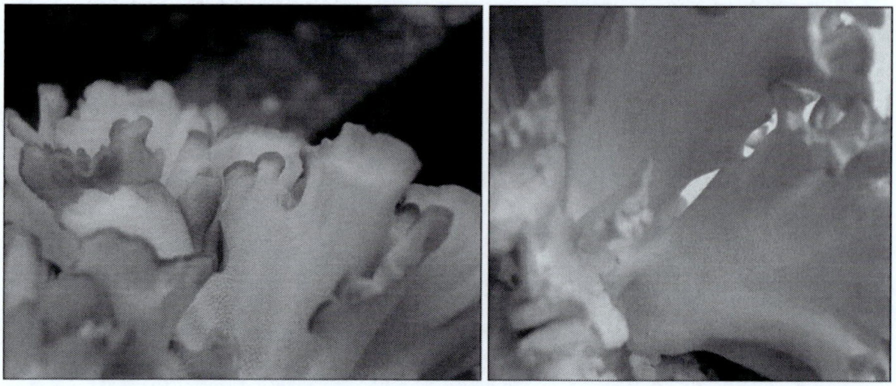

Figure 32. Pores of *Grifola frondosa* fruiting bodies.

Methods for the Solid-State Cultivation of *G. frondosa* at the Pilot Scale

G. frondosa develops at low temperatures in the presence of light and high air concentrations. It is considered a white-rot saprophyte. The environmental conditions for the growth and development of Maitake mushrooms are described in Table 26. These conditions were determined experimentally in China for the proper growth and development of this mushroom. In particular, Wu et al. (1997) found that this species is very demanding in terms of aeration requirements. Thus, in cultivation rooms, the air exchange should not be less than five or six times per day, as poor aeration conditions produce malformed or atrophied fruiting bodies (Chang and Miles 2004).

Table 26. Growth parameters of *Grifola frondosa*

Parameters	Mycelial growth		Primordia formation	Fruiting development	
	Range	Optimum		Range	Optimum
Temperature (°C)	5–32	20–25	18–22	10–25	15–20
Relative humidity (%)	60–70		80–90	80–95	85–95
Light intensity (luxes)	50		200	200–500	
pH	5,5–6,5		-	-	

Source: Wu et al. (1997).

Table 27. Maitake synthetic-log cultivation in Japan

Parameter	Spawn run		Primordia initiation (in sealed bags[a])	Fruiting body development (in cut-open bags)
	Stage I	Stage II		
Temperature (°C)[b]	22–23 (air); 24–25 (in substrate)	22–23 (air); 24–25 (in substrate)	22 – 23 (23 best for differentiation)	16–18
Relative humidity (%)	70	60-80	Same as spawn run	85–95
Light intensity (luxes)	N/A	50	50	200–500
CO_2 ventilation	Mycelia tolerate higher CO_2	Mycelia tolerate higher CO_2	< 0.30% (3000 ppm) Bags 3-4 cm apart to allow ventilation	0.10% (1000 ppm) 0.15% (1500 ppm) results in small/thin caps; after 2-3 days, cut bag on top for ventilation
Time (days)	30–35	40	7	20-25

[a] Heat-sealed bags with microfilters for breathing.

[b] Fluctuations in the temperatures should be minimized to avoid condensation.

Source: Huang (1997).

For the artificial cultivation of *G. frondosa*, researchers such as Chen et al. (1999) isolated strains of the fungus from the wild, took them to the laboratory and adapted them for development in artificial culture. These strains were selected to search for the best performance during in vitro culture. The strains were incubated on malt extract agar (MEA) for 10 days at 25°C under penumbra. In Japan, Huang (1997) developed the artificial solid-state culture conditions required for producing *G. frondosa*, as summarized in Table 27. In this table, the conditions required to obtain fruiting bodies within approximately 105 days are described when using heat-sealed bags with microfilters for gas exchange. Stamets (1993) also determined the growth and development conditions for Maitake production. This author recommends a single temperature for the incubation stage (between 22°C and 24°C); for primordia induction, a thermal shock at 10-15°C for 5-10 days is suggested, followed by maintaining a low temperature for another 10-15 days once the primordia have been formed. Subsequently, the temperature can increase to 13-15°C for the growth of the fruiting bodies.

Regarding the relative humidity (RH), it should be maintained between 90% and 95% during all growth stages with the exception of the later one (corresponding to the growth of the fruiting bodies), for which 85-90% RH is proposed.

Production of *G. frondosa* Inoculum

For the production of *G. frondosa* inoculum, it is necessary to follow several stages that guarantee the quality of the inoculum obtained for the efficient production of the fruiting bodies of this mushroom.

Strain Selection and Maintenance

The most common culture media for the growth and development of Maitake mushrooms are potato dextrose agar (PDA) and malt extract agar (MEA), as for *G. lucidum* (see Chapter 2). In Table 9, alternative culture media for the mycelial vegetative growth of macromycetes are shown; these media may also be used for the massification of *G. frondosa* biomass.

Maitake strains are obtained from the wild. These strains then undergo vegetative reproduction and adaptation in selective media for their growth. Alternatively, the strains can be obtained from specialized laboratories. To obtain strains from the wild, the specimens are collected, transported on paraffined paper, stored under refrigeration and induced to engage in vegetative reproduction, as with Reishi mushroom strains. For the vegetative reproduction of a Maitake specimen, a petal from the parts that make up the carpophore is selected, and a piece is cut and separated as soon as possible from the tubules in which the spores are formed. Subsequently, the pieces of carpophore are washed in chlorinated water with 1-2% sodium hypochlorite for a maximum of 5 min, and then they are washed in sterile distilled water.

The washed mushroom fragments are inoculated on PDA, MEA, or other culture medium indicated for fungal growth in slant test tubes. The inoculated tubes are incubated at 25°C under penumbra for 10-20 days or until the colonization of the slant agar surface.

The biochemical tests discussed in Chapter 2 are also applied for selection and viability determinations of *G. frondosa* strains, particularly for semiquantitative tests of laccase and cellulase enzymatic activities. Once the results of the biochemical tests are obtained, the mycelium is massified to generate the amount of mycelial biomass required to produce commercial spawn or a final inoculum. In the following research case, the cultivation of *G. frondosa* strains on synthetic media during the vegetative phase is explained in detail as an account of the research performed at the Bioprocess and Agro-industry Plant at the University of Caldas.

Research Case 14. Cultivation of Grifola frondosa Strains in Vegetative Phase on Agar-Based Media

Determining the Viability of G. frondosa Strains

In the Bioprocess and Agro-industry Plant at the Universidad de Caldas (Manizales, Colombia), the first protocol in the viability test of a *G. frondosa* strain purchased from a strain bank located outside of Colombia was developed. The first test performed on this strain consisted of determining its growth and development viability during the first five generations of mycelium propagated on PDA. For this purpose, the initial purchased strain (F_0) was sown on PDA and incubated for the mycelium to grow and develop. From the corresponding Petri dish, a piece of mycelium was cut and sown onto another plate containing PDA (F_1); this procedure was successively repeated until the fifth generation (F_5). In Figure 33, the mycelial growth of *G. frondosa* mushrooms is observed over several generations. These tests were performed to determine whether there was an effect on the growth rate of the mycelium with the successive generational passages.

There were no significant differences in the growth rates of the biomass on PDA until the F_5 generation. In addition, cultivation tests were performed to produce fruiting bodies by using the mycelial material obtained from each one of the generations as starting inoculum without detecting the important changes in the biological efficiency (*BE*) measured from the substrates under study.

According to the obtained results, a viability test protocol for *G. frondosa* strains was developed not only for the strains purchased from international strain banks but also for those stored in the strain bank of the Bioprocess and Agro-industry Plant at the Universidad de Caldas. These tests allow for the determination of the strain quality in terms of their ability to grow, growth rate, and expression of the enzymes characteristic of the organism, among others. In Figure 34, the scheme of the protocol developed for the viability determination of *G. frondosa* strains is shown. The test culture media are PDA and MEA. The steps to perform for the Maitake strain viability determination are as follows:

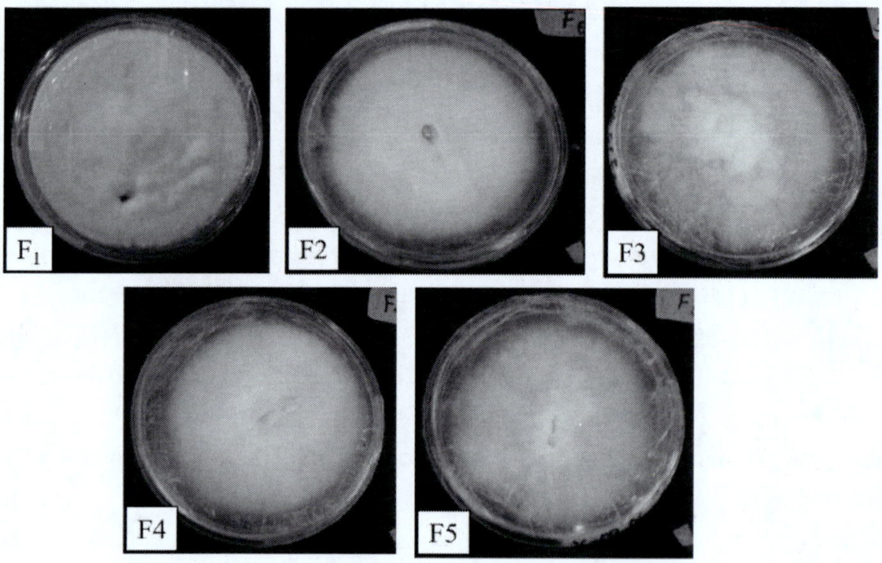

Figure 33. Growth of *Grifola frondosa* mycelium on potato dextrose agar (PDA) over five generations in a strain purchased from a strain bank.

- Reception of the purchased strains or of the extended *G. frondosa* carpophore tissue and temporary storage under refrigeration at 3-4°C.
- Preparation of the PDA or MEA media on which the strains will be inoculated. If both media are employed, the same number of Petri

dishes with identical diameters should be inoculated to compare the fungal growth on both culture media.

- Petri dishes must be washed and sterilized ahead of time. Next, a Cartesian plane is drawn on the outer face of the base of each Petri dish with an indelible marker and a ruler; all the Petri dishes should be numbcred. At least five Petri dishes should be prepared per culture and per strain.
- The culture medium is poured hot in a laminar flow cabinet and allowed to solidify.
- The culture medium in the Petri dishes is inoculated in the center of the plate, ensuring that it is made at the zero point of the Cartesian plane (see Figure 35).
- Petri dishes are incubated in an incubator or oven where the culture conditions may be set at 25°C with a dry environment (approximately 40% relative humidity).
- The growth diameter should be measured from the inoculation point. The value of this diameter must be determined on the two axes of the Cartesian plane as previously drawn, that is, measuring on the *x*-axis and on the *y*-axis and then registered. This measurement should be obtained every day until the first Petri dish reaches full colonization; this measurement corresponds to a diameter of 90 mm, but it is possible to accept a 89 mm mycelium diameter in one of the axes of the Cartesian plane.
- When the first Petri dish presents a state of complete colonization, the incubation stage is considered finished. For viable and intact *G. frondosa* strains, the incubation time on PDA is approximately 15-25 days, and on MEA, it takes 25-30 days.
- With the growth diameter of each plate and the time of colonization, the growth rate of the mycelium (in mm/day) is calculated. All the Petri dish cultures that have reached a diameter of approximately 72 mm on one of the axes are selected; the other plates should be discarded.

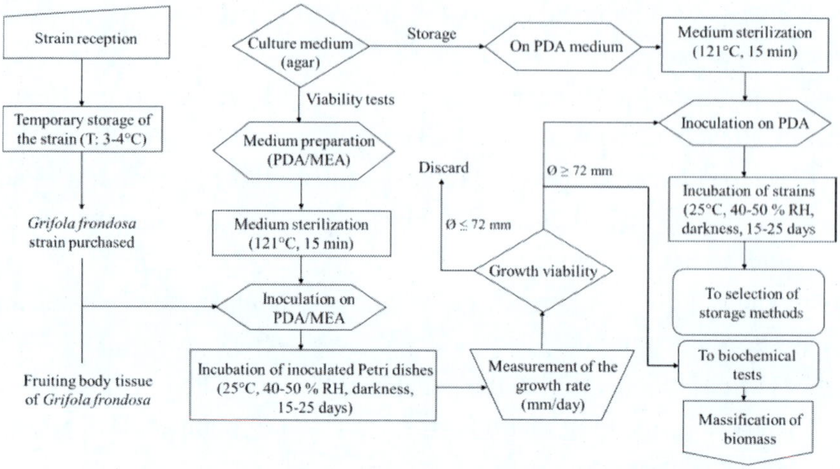

Figure 34. Protocol for determining the viability of *Grifola frondosa* strains in the Bioprocess and Agro-industry Plant at the Universidad de Caldas.

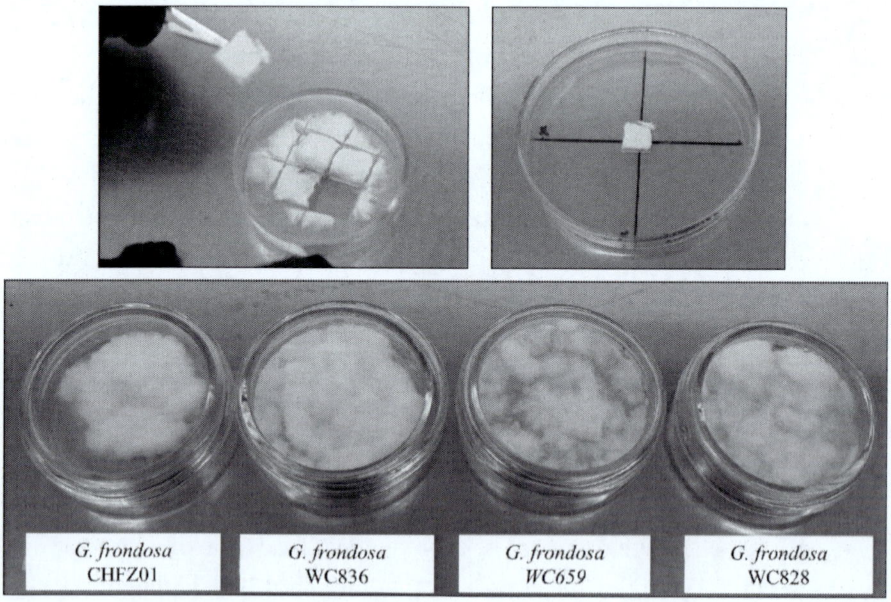

Figure 35. Inoculation of *Grifola frondosa* mycelium at the zero point of the Cartesian plane in a plate to determine the viability of the strain and to measure its growth rate.

- Petri dishes whose mycelium have reached a colonization of 80% or more are collected for biochemical tests, which consist of qualitative or semiquantitative tests of laccase and cellulase production (endoglucanases and exoglucanases). The results of these tests for Maitake mushrooms are presented in Research Case 3.

- Once it is verified that the strains produce the enzymes needed for solid-state *G. frondosa* cultivation, one portion of the plates is employed for strain storage and maintenance, and the other one is used for biomass massification with the purpose of producing spawn and, subsequently, the fruiting bodies of the mushroom.

Tissue Extension from G. frondosa Basidiomes

The renewal of a *G. frondosa* strain may be accomplished by considering the selection of a fruiting body that meets the requirements of the market. The tissue extension procedure from Maitake carpophores is the same as that described in Research Case 2 (see Chapter 2). Figure 36 shows the procedures for renewing working *G. frondosa* strains from the extension of tissue in the laboratory.

Research Case 15. Biochemical Tests for Screening Lignocellulolytic Enzymes as a Measure of the Lignocellulosic Material Degradation Capacity of G. frondosa

In Research Case 3 (Chapter 2), the main results obtained from the degradation study performed in a previous work (Montoya, Sánchez, and Levin 2014) on ten white rot fungal species were presented. *G. frondosa* was one of the ten species to be assessed.

The average decolorization halos determined during cellulose degradation tests after being revealed with Congo red for all the evaluated species are indicated in Table 10 (Chapter 2), while the results of the tests on the ability of these same fungal species to react with two specific substrates that are indicators of laccase and lignin peroxidase production are shown in Table 11. In this section, the outcomes obtained for *G. frondosa* are disclosed (Montoya 2012). The studied strain of this mushroom generated a cellulose degradation halo of 10 mm (from an average of three repetitions in 30 inoculated Petri dishes; see Table 10). These degradation halos for carboxymethyl cellulose and crystalline cellulose by the action of *G. frondosa* are shown in Figure 37. Considering the data, it is clear that the assessed Maitake strain has a lower cellulose degradation ability than the Reishi mushroom strain; nevertheless, this result implies that the Maitake has important potential to degrade and consume the different cellulosic materials contained in the substrates used for its production.

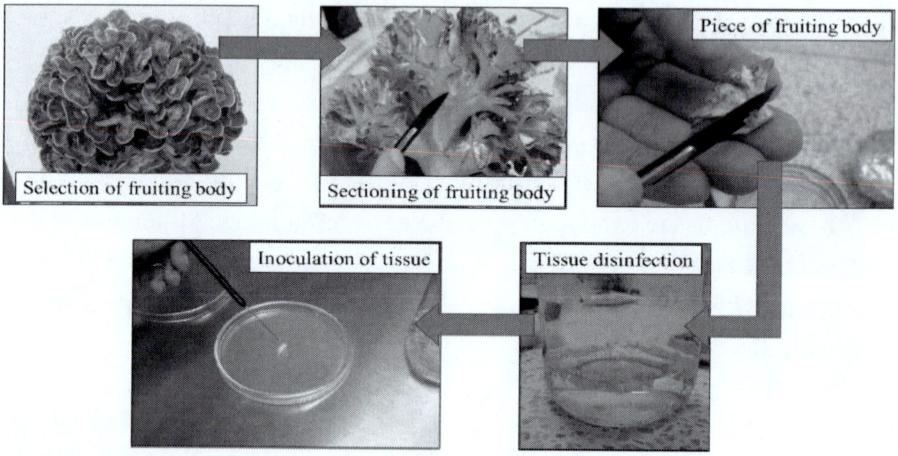

Figure 36. Overall procedure for renewing *Grifola frondosa* strains through tissue extension from their fruiting bodies in the Bioprocess and Agroindustry Plant at the Universidad de Caldas.

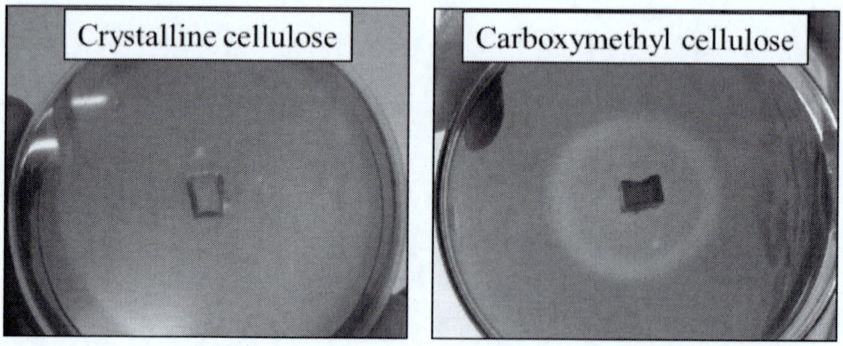

(a) (b)

Figure 37. Formation of decolorization halos during the degradation test on two cellulose forms by *Grifola frondosa*. (a) Carboxymethyl cellulose. (b) Crystalline cellulose. Source: Montoya (2012).

The formation of green halos in ABTS-containing solid media inoculated with the Maitake strain is presented in Figure 38. The reaction with ABTS indicates that the mushroom is a white rot fungus since, in the presence of this compound, the mushroom excretes laccase, which catalyzes oxidation reactions and forms green pigments with subsequent pigmentation from yellowish to reddish tones as the successive reactions proceed. From the results indicated in Table 11, it is evident that *G. frondosa* has a moderate ability to produce laccase and lignin peroxidase. Thus, different waste materials containing lignin may be employed by the studied *G. frondosa* strain.

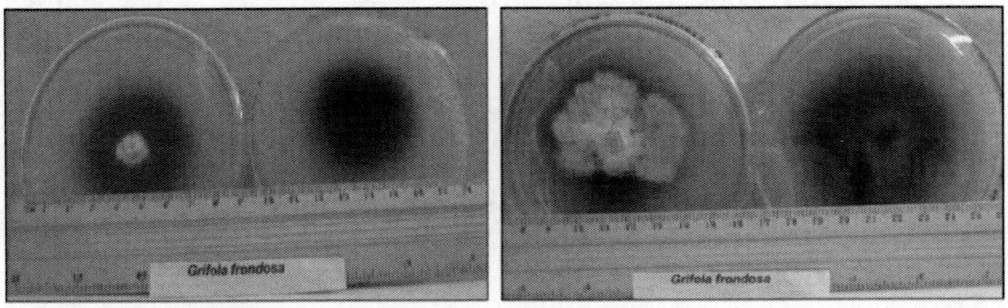

(a) (b)

Figure 38. Green halo formation by *Grifola frondosa* on a medium containing ABTS. (a) Pigmentation on the day of inoculation. (b) Pigmentation of the medium at 14 days of incubation. Source: Montoya (2012).

Conservation of Grifola frondosa Strains

A great variety of methods for conserving macromycetes strains has been developed in the past few years. These methods are specific to each fungal species under conservation. Kirsop and Snell (1984) and Smith (1993) developed conservation methods for many microorganisms and fungi. In addition, researchers such as San Antonio and Hwang (1970), San Antonio (1979), and Jong (1978) made enormous contributions to selecting macromycete conservation methods according to the characteristics of each species. Thus, the conservation methods should be selected by factors such as the necessary and available equipment needed and the specific requirements of each species. The most commonly used conservation methods presented in Chang and Miles (2004) are briefly described below:

Short-Term Storage

Short-term storage is a traditional method for macromycete conservation through periodic transfers. The cultures are always available for subcultures. The success of this storage method relies on three factors: the frequency of the transfer, the maintenance medium, and the storage temperature. The period between subcultures depends on the species. This method is considered useful for periods ranging from 3 up to 12 months. There are two main variants for conserving macromycetes: culture practices and small flat-sided culture bottles.

Culture practices are a method of subculturing with transfers every six months using Petri dishes at a storage temperature of 5°C. The strains that are stored and maintained by periodic transfer should be checked constantly to discard the ones that exhibit dehydrated, hard, or strange-looking mycelium. In addition, strains of species grown at temperatures above 30°C should be transferred more frequently (every two months) and stored at room temperature.

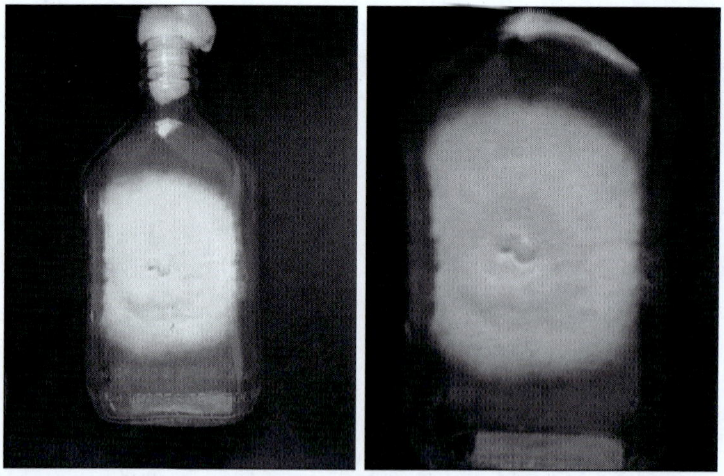

Figure 39. *Grifola frondosa* mycelium spread in a flat bottle.

For the second variant, the strains are conserved in small, flat-sided culture bottles, and they are subject to periodic transfers. This variant is often used as a maintenance method for propagation cultures. These bottles have a capacity of 20 mL of culture medium. The medium should be solidified with the bottle in a horizontal position; subsequently, a piece of fungal mycelium is inoculated and incubated until the mycelium grows as observed in Figure 39.

Long-Term Storage

The purpose of long-term storage methods is to slow cellular metabolism. This aim can be achieved by combining different methodologies, such as the ones described below.

- Nutrient starvation. Storage in distilled water suppresses the growth of the mycelial mass and has also been attributed to the suppression of pleomorphic changes and the inhibition of genetic changes in fungi. This methodology has been reported as a conservation alternative, but it has not been well explored by the authors of this book.
- Oxygen limitation. This technique consists of limiting the concentration of oxygen available to the fungal strains to slow down their metabolism. To this end, the tissue is inoculated in a slant tube with agar, and sterile mineral oil is added; the slant tube is stored under refrigeration at 5°C. This method prevents the passage of oxygen to the mycelium and therefore retards the metabolism of the fungi.
- Lyophilization. This method is widely used for the conservation of fungal spores, but not mycelium. The freezing and subsequent removal of water by sublimation does not cause apparent damage to the spores; on the contrary, it has been shown in several investigations that the spores of many macromycetes species are well preserved for long periods of time by lyophilization.
- Freezing. This method involves the freezing of the mycelium of different fungal species in liquid nitrogen. The metabolism of the mycelium decreases with the decrease in temperature, and when the temperature is lower than -130°C, it practically stops. Therefore, periodic transfers of the mycelium are not required, and in theory,

the mycelium can survive in storage indefinitely. The procedure involves mycelium propagation on proper agar, such as PDA, and then very small pieces, measuring 0.5 to 0.3 mm in diameter or per side, are cut. Subsequently, these pieces are transferred to vials to which a cryoprotectant is added (e.g., 10% glycerol solution or an aqueous solution of dimethyl sulfoxide). The vials can be packed in aluminum foil in groups or they can be sealed with paraffin. The vials are then placed in the compartments of a thermos and submerged in nitrogen. To recover the mycelium, the vials are removed from the thermos and submerged in a water bath at 30°C for 10 min. Then, they are washed with 75% alcohol for disinfection with subsequent drying. The mycelium of the vials is transferred to a new culture medium in a Petri dish and then incubated at 25-30°C until the plate is 90% colonized.

Stamets (2000) proposed using the slant tube technique to conserve of the strains of many macromycetes, including *G. frondosa*. For this purpose, 100×20 mm test tubes with caps are used. PDA or MEA culture medium is prepared, sterilized at 121°C for 15 min and then dispensed into test tubes and solidified at a 15° or 20° slant angle. A piece of tissue is then inoculated and incubated at 25°C under penumbra until the complete colonization of the slant agar surface. The test tubes are stored in boxes with lids that are equipped with gauze and cotton and refrigerated at 1-2°C. In this way, the strains can be stored for months and can be used as working material. This author proposes other conservation methods for macromycete strains that have already been described.

Research Case 16. Protocols for the Maintenance and Conservation of G. frondosa Strains

For the maintenance and conservation of *G. frondosa* strains, the methodologies described in Research Case 4 of Chapter 2 are applied with some modifications. The first method that was successfully applied to Maitake mushroom consists of conservation in Eppendorf tubes containing the strain on PDA medium with sterile distilled water, which are then closed with paraffin and stored at room temperature (17°C). This conservation method has shown strain viabilities of up to 90% over two years, even with successive passes for growth tests and determinations of the main enzymatic activities every third month.

The second methodology consists of the preservation of a piece of fungal mycelium spread on PDA agar in Eppendorf tubes with 10% glycerol that are frozen at -20°C. This technique has been found to yield strain viabilities of up to 90% for six years, with successful transfers for growth tests and determinations of the main enzymatic activities every six months. The third method used for the Maitake consists of periodic transfer of the mycelium conserved in slant tubes under refrigeration at 1-4°C. A transfer is performed every two months. This technique is used for the maintenance of the active working strains. The *G. frondosa* strains under storage are shown in Figure 40.

Figure 40. Conservation of *Grifola frondosa* strains in the Bioprocess and Agro-industry Plant at the Universidad de Caldas.

Massification of the G. frondosa Mycelial Biomass

The massification of *G. frondosa* mycelial biomass is initiated once the selected strain has passed the viability (see Figure 35) and biochemical tests (see Research Case 15). The overall scheme for the massification of *G.*

frondosa mycelial biomass is depicted in Figure 13. The starting point in that scheme is the lineage strain, i.e., the strain provided by the supplying laboratory, such as international strain banks. The procedure for biomass massification was disclosed in Research Case 5 (see Chapter 2). This massification consists of reproducing the mycelial biomass successively from well-developed and fresh *G. frondosa* mycelium through successive passes of the mushroom mycelium onto PDA or MEA media, as illustrated in Figure 33. As in the case of *G. lucidum*, it is necessary to calculate how many 90 mm Petri dishes with culture medium colonized by the working Maitake strain are required, considering that one of these one Petri dish can be divided into 24 to 30 parts. Figure 41 illustrates how the mycelium of *G. frondosa* should be divided in the colonized Petri dish. It is worth highlighting that the Maitake mycelium is white and cottony when it grows on agar and not rhizomorphic. The mycelium grows unevenly, without forming the circular colonies typical of most fungi. Thus, some regions of the mycelium exhibit a higher growth rate than others. For example, if a Cartesian plane is drawn on the Petri dish, the mushroom grows faster across one axis than the other, i.e., the growth of this mushroom is longitudinally linear. This growth pattern seems to be a characteristic of this species, as observed in the 10 study strains that are currently stored in the Bioprocess and Agro-industry Plant. The mycelium develops light brown tones along the outer peripheral edges when it ages. At maturity, the dense mycelial layer can be detached directly from the agar.

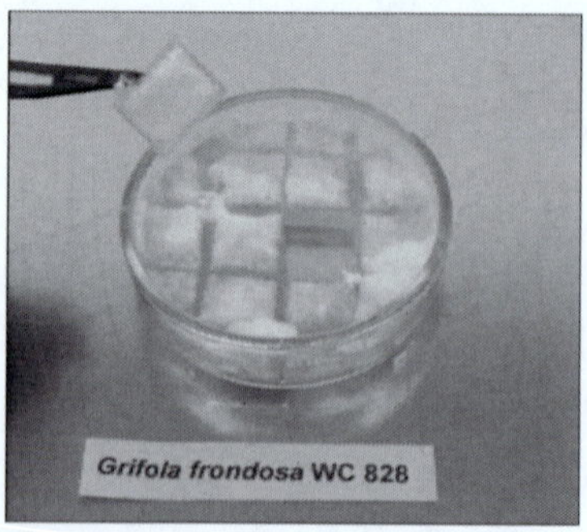

Figure 41. Division of *Grifola frondosa* mycelium in a colonized Petri dish.

Production of G. frondosa Spawn

The production of macromycete spawn for producing *G. frondosa* mushrooms is performed from the exponential expansion of mycelial biomass on cereal grains or lignocellulosic waste as described in Chapter 2. In that chapter, it was disclosed that each mushroom species exhibits preferences for one or another cereal grain. Different procedures for *G. frondosa* mushrooms have been developed at the Universidad de Caldas, and they range from the production of the inoculum for synthetic cultivation to the production of its fruiting bodies at different scales. In the following research case, the assessment of *G. frondosa* inoculum production on four types of cereal grain is discussed to define the best conditions for its growth and development.

Research Case 17. Development of Protocols for Producing G. frondosa Spawn

In the Bioprocess and Agroindustry Plant at the Universidad de Caldas, the protocol for producing *G. frondosa* mushroom spawn has been developed, as presented in a previous study (Montoya, Varón, and Levin 2008). The procedure developed to produce the primary generation spawn (also called the mother spawn) and second-generation spawn (also called the production inoculum, commercial spawn or, simply, spawn) of *G. frondosa* mushrooms is depicted in Figure 42.

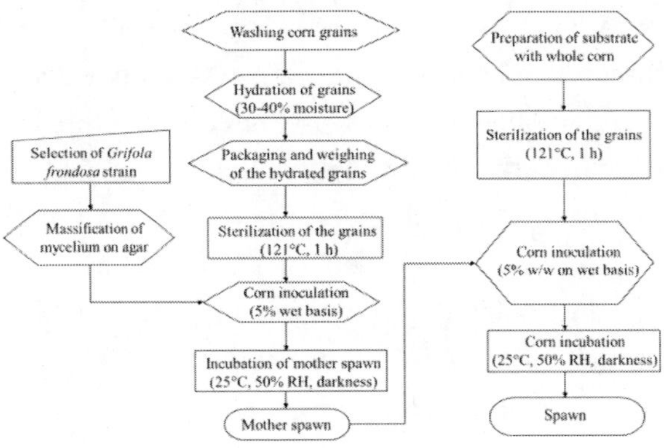

Figure 42. Scheme for producing the mother and commercial spawn of *Grifola* frondosa in the Bioprocess and Agro-industry Plant. RH: Relative humidity.

Production of Mother Spawn

During the first stage of research, the procedures for producing the mother spawn of this mushroom on four different cereals grains were defined and assessed. The whole grains on which the tests were performed were rice, red sorghum, barley, and corn.

The objective of this study was to determine which of the four grains offers the best conditions for the growth and development of *G. frondosa* strains in the vegetative phase. The cereal grains were hydrated to 33-40% moisture, and then 150 grams of each hydrated grain was packed in 500-mL bottles with a hole in the cap, which utilized cotton as a filter for gas exchange. The bottles containing hydrated cereal were sterilized at 121°C for 1 h. The cereals in the flasks were inoculated with five square pieces measuring 1 cm on the side of the fungal mycelium spread on agar, and they were incubated at 25°C under penumbra. This stage is needed to adapt the mushroom mycelium to each cereal grain to obtain the *G. frondosa* mother strain. The best behavior in terms of growth corresponded to the mushroom inoculated on corn, for which the complete colonization of the grains was achieved within 21 days (see Figure 45a). The slowest growth rate corresponded to the mushrooms inoculated on barley grains (29 days of incubation, see Figure 45b).

Production of Second-Generation Spawn

To produce commercial spawn, the substrate is also prepared on corn grains. The procedure for washing the grains consists of washing with clean water until the impurities, broken grains, leaves, and powders are removed. After the grains are washed, they are hydrated in boiling hot water until reaching 30-40% moisture. Subsequently, the grains are cooled to room temperature to be packed in 500-g bioriented polypropylene bags. The substrate packed in the bags is sterilized for 1 h. Once the sterilized substrate has cooled, it is used for inoculation in the laminar flow cabinet; the mother spawn is used as inoculum at 5% (w/w) on a wet substrate basis, which corresponds to 25 g of inoculum for every 500-g bag of sterilized grain. The inoculum is distributed homogeneously throughout the material. The bags containing inoculated material have a PVC ring and a sterilized cotton plug as a filter for air exchange or a filtering cloth on the top of the bag (see Figure 43). The bags are then placed in an incubation room at 25°C, 50% relative humidity, under penumbra, and with an air exchange to vent for 15 min per day. The incubation of the material begins with medium colonization from each inoculation point, that is, each cereal grain is colonized by the mycelium of the fungus. The incubation period of the second-generation spawn is 25 days for the complete colonization and thickening of the mycelium. During this period, the bags should not be moved, because *G. frondosa* has no capacity to recover from mycelial breakage. The mycelium turns white and makes soft threads through the corn grains, which are not very dense and are sometimes yellowish. The spawn is stored under refrigeration at a 3-4°C maximum temperature for one month.

Production of Fruiting Bodies at Small and Pilot Scales

The commercial cultivation of *G. frondosa* is recent and relatively rare. The first commercial Maitake production was reported in Japan in 1981. In Colombia, the cultivation of *G. frondosa* was developed at the Universidad de Caldas from 2006 by the authors of this book, and there are no additional reports of Maitake culture in the country at any scale. In the Bioprocess and Agro-industry Plant, protocols for producing Maitake fruiting bodies have been developed from small-scale cultivation up to the commercial scale.

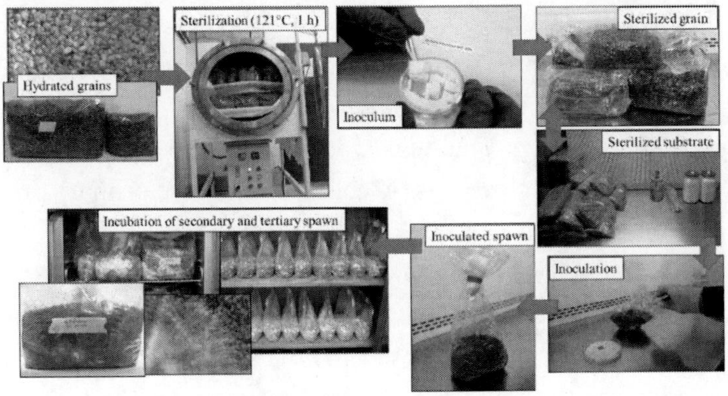

Figure 43. Procedure for producing secondary and tertiary *Grifola frondosa* spawn at the Bioprocess and Agro-industrial Plant (Universidad de Caldas).

In the following research cases, two examples of how the production of *G. frondosa* fruiting bodies was developed in the Plant and in the former Laboratory of Macromycetes at the Universidad de Caldas are briefly described.

Research Case 18. Effect of Culture Parameters on Maitake Production at a Small Scale

In a previous study (Montoya, Varón, and Levin 2008), the solid-state cultivation conditions for producing *G. frondosa* at a small scale were established. "Small scale" indicates the processing of 20 kg of substrate packed into 1-kg bags. Experimentation was performed using a strain of *Grifola frondosa* (PSUMCC 922) obtained from the Pennsylvania State University Mushroom Culture Collection (USA). Two substrate formulations based on oak sawdust, coffee spent-ground, and corn bran and supplemented with calcium carbonate and commercial sugar were employed. The substrate was packed in bags, sterilized, inoculated, and incubated under controlled conditions. The first formulation tested was as follows (on dry basis): 75% oak sawdust, 23% corn bran, 1% sucrose, and 1% calcium carbonate. In the second formulation, all the oak sawdust content was replaced by 50% oak sawdust and 25% coffee spent-ground; the other components had the same concentrations as those of the first formulation. Water was added to the substrates to reach 58-60% moisture, and then they were packed in bioriented polypropylene bags measuring 32 cm high by 12 cm in diameter, with a 2.54 cm^2 filtering cloth at their tops for gas exchange. Each bag was filled with 1 kg of moist substrate and sterilized at 121°C for 1 h. The bags were then inoculated with 3% spawn on a wet basis according to the mass of the substrate. The primary processing steps are depicted in Figure 44.

The environmental conditions defined in this research case for the growth and development of Maitake mushrooms through solid-state cultivation at a small scale are presented in Table 28. The environmental parameters identified as the most relevant corresponded to the two subphases of the mycelial growth (spawn run) stage, i.e., the vegetative phase of the mushroom. Thus, during the first subphase, the temperature was set to 25°C during the first week and to 20-21°C during the following 28 days of colonization under penumbra without air exchange. During the second subphase, the temperature was maintained at 18-20°C under penumbra without air exchange for 40-45 days until acquiring an orange pigmentation (this color indicates that the substrate is ready for primordia formation, as shown in Figure 45d).

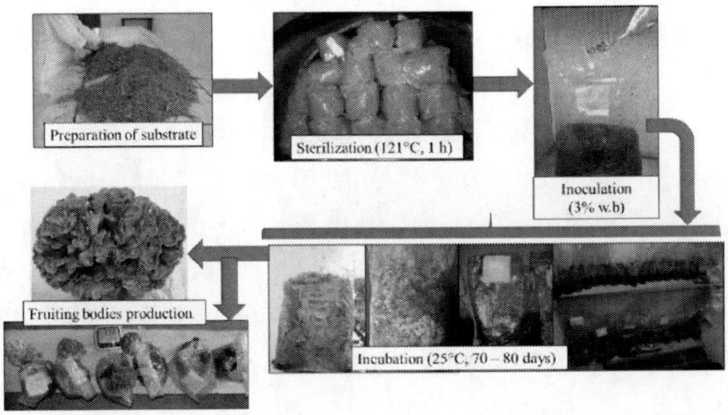

Figure 44. Small-scale *Grifola frondosa* production cycle by solid-state cultivation.

Table 28. Growth parameters defined for *Grifola frondosa* on synthetic solid-state cultivation at a small scale in Manizales (Colombia)

Parameter	Mycelial growth (in sealed bags)		Primordia formation (in sealed bags)		Fruiting development (in cut-opened bags)
	Subphase I (until complete colonization)	Subphase II (until exudate formation)	Induction	Formation	
Temperature (°C)	25 (1st week), then 20-21	18–20	10	16–18	16–18
Relative humidity (%)	60–65	60–65	60–65	70–80	70–80
Light intensity (luxes)	Not required	Not required	Not required	50–100	50–100
Ventilation	Not required	Not required	Not required	One air exchange (10 min) every 12 h	One air exchange (10 min) every 2 h
Time (days)	30–35	40–45	1	7–10	3–5

To induce the primordia, a thermal shock of 10°C was applied to the compacted substrates with orange pigmentation and exudates for 24 h. Subsequently, the substrates were placed back in the culture room at 16-18°C, 70-80% relative humidity, and 50-100 luxes with 10-min air exchanges every 12 h for 10 days. Finally, the growth and development of the fruiting bodies up to the point of harvest was performed under the same conditions as those of primordia formation for up to 5 days, except that the air exchanges were increased to 10 min every 2 h.

The different stages for producing Maitake fruiting bodies from spawn running until the harvest of carpophores from oak-sawdust-based substrate are illustrated in Figure 45. Picture (c) in that figure shows the spawn run after 15 days with undifferentiated white mycelia, while Picture (d) shows the mycelial growth after 75 days with orange-brown exudate. Then, a mycelial mat (or mycelial coat) on the substrate surface is formed as presented in Picture (e); the dark gray exudate indicates the place where primordia will develop, as shown in Picture (f), where an uneven topography on mycelial surface is observed. Picture (g) presents the so-called brain state: as the dark grayish black primordia grow, convoluted folds appear on the surface, resembling a brain. Picture (h) shows the post-brain stage: further growth includes the unfolding of the convoluted folds on the surface of the dark primordia into overlapping young pilei (caps) that have formed in a cluster; this step is followed by the elongation of the lateral stems, each with a young pileus on the upper portion. Picture (i) illustrates the cauliflower stage; the developing mushroom cluster (lighter, almost white) has overlapping petals with elongated lateral stems, each of which has a young cap on the upper portion resembling a cauliflower. Picture (j) shows the fruiting body ready for harvest and Picture (k) corresponds to an overmature carpophore. Picture (l) presents harvested fruiting bodies; the Maitake mushroom progresses from grayish brown when young to light gray, grayish white, or light brownish yellow when older.

In this research, the substrates formulated from coffee spent-ground did not fructify and therefore did not produce Maitake fruiting bodies. In particular, the zinc content of the substrate containing coffee spent-ground had a strong effect on the yield of the mushrooms (see Research Case 24 in

Chapter 7). Therefore, the substrates that fructified were those with the formulation based on oak sawdust, with an average biological efficiency of 35.3% (see equation 1).

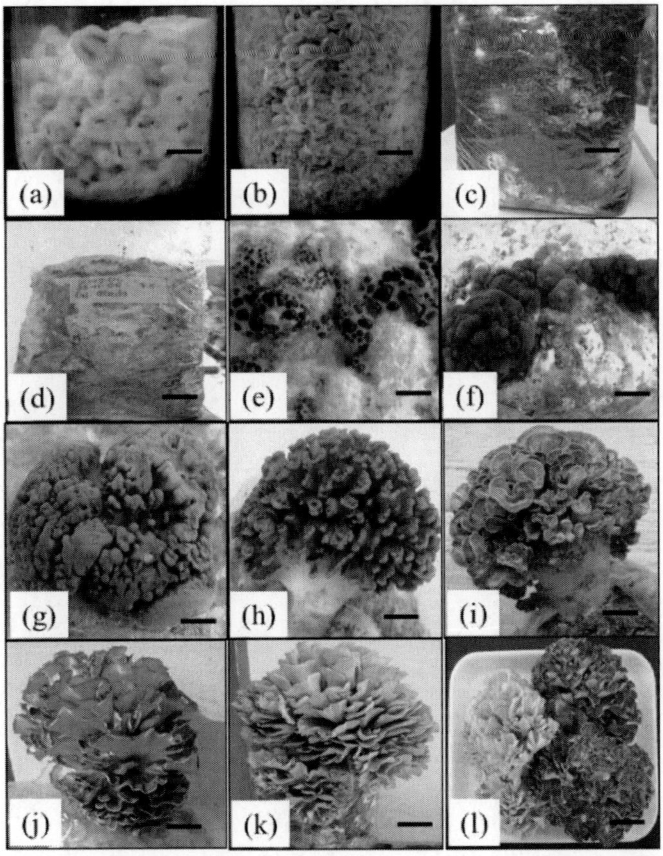

Figure 45. Growing stages of *Grifola frondosa* mushrooms cultivated in the Bioprocess and Agro-industry Plant at the Universidad de Caldas. Mother spawn obtained from corn (a) and barley (b). Mycelium development in a substrate based on oak sawdust and corn bran: White mycelium undifferentiated (c), orange-brown exudate (d), dark gray exudate (e). Fructification: Unequal topography in the formation of the primordia (f), brain state (g), post-brain state (h), cauliflower state (i), state of the cauliflower cluster (j), and mature mushroom (k). Collected mushrooms (l), with an old mushroom on the left and a young mushroom on the right. The black bar represents 1 cm in length. Source: Montoya, Varón, and Levin (2008).

This *BE* and the quality of the carpophores obtained in this research case were comparable to those obtained by other researchers. Shen and Royse (2002) developed a scale from 1 to 4 to determine the quality of Maitake fruiting bodies, in which 1 corresponds to the highest quality and 4 to the lowest. In this research case, carpophores with a quality of 1 were obtained according to the guidance of these authors, presenting dark gray basidiomes with well-formed bodies and a firm texture.

Research Case 19. Development of G. frondosa Culture at the Pilot Scale

The cultivation of Maitake mushrooms at the pilot scale was developed in the Bioprocess and Agro-industry Plant at the Universidad de Caldas in Manizales (Colombia) at 2,250 m above sea level with an average temperature of 17°C. This culture was implemented according to the protocol established by Montoya, Varón, and Levin (2008) for tropical weather. The pilot-scale cultivation was applied to four *G. frondosa* strains: WC 836, 828, and 659 obtained from the Pennsylvania State University Mushroom Culture Collection (USA) and CHFZ01 from the Fujian Agriculture and Forestry University (China).

Several aspects of the protocol for small-scale solid-state cultivation (see Research Case 18) were modified when the production level was increased. The following modifications are highlighted:

- Increase in the size of the substrate unit from 1 kg to 2 and 3 kg mix of solid materials. This increase was performed to define a suitable production size for implementation of the culture at higher scales.
- With the increase in the size of the substrate unit, it was necessary to increase the sterilization time from 1 to 2 h at 121°C through the Tyndall method, which consists of subjecting the substrate units to double sterilization.
- The thermal shock time at 10°C was increased from 24 h for bags with 1 kg substrate up to 60 h for bags with 2 and 3 kg substrate.

The environmental conditions for the pilot-scale cultivation were not substantially modified from the conditions established by Montoya, Varón, and Levin (2008). The biggest modification corresponded to the times of each stage (see Table 29). For example, during the second stage of incubation corresponding to the formation of the mycelial cover, the compaction of the substrates and their pigmentation with the formation of exudates increased from 45 to 55 days, or 10 days more than for the small-scale culture. Similarly, the induction and formation of the primordia until the harvest time lasted up to 20 days for the bags containing 3 kg of substrate. Additionally, a test was performed to define the thermal shock temperature of the bags containing 2 and 3 kg substrate. For this purpose, half of the substrate units of each size were subjected to thermal shocks in a cold room for 36 h to lower the temperature of the substrates down to 10°C; for the other half, the temperature in cultivation room was reduced down to 15-16°C for two days. It was concluded that it is sufficient to lower the temperature of the substrates incubated with *G. frondosa* mushrooms down to 15-16°C to induce the fruiting bodies at scales greater than that obtained in the laboratory. In addition, the ventilation of the culture rooms at each stage was modified during the incubation by performing two air exchanges of 1 h every 12 h; during the primordia formation and fructification stages, the air exchanges were increased from 10 min every 12 h to 2 h every 12 h. The remaining conditions (temperature, relative humidity, and light intensity) were very similar to those of the solid-state cultivation at a small scale.

The yield in terms of the *BE* of the four strains evaluated for each one of the tested substrate sizes is summarized in Table 30, where each value corresponds to the average of 10 replications for the productive bags; there were 10 batches of 200 kg substrate, of which 100 kg was packed into bags containing 2 kg substrate each and the other 100 kg was packed into bags containing 3 kg substrate each. The productive bags were those that successfully fructified. The lowest *BE* values were always obtained in the units containing 3 kg of substrate, in which a *BE* of 12% was obtained for strain WC836.

Table 29. Environmental conditions for *Grifola frondosa* under synthetic solid-state cultivation at the pilot scale in Manizales (Colombia)

Parameter	Incubation		Primordia formation		Fruiting development
	Subphase I	Subphase II	Induction	Formation	
Temperature (°C)	20–25	20	10 15–16	16–18	16–18
Relative humidity (%)	50–60	60	60	70–75	80–85
Light intensity (luxes)	Not required	Not required	50–150	100–150	100–150
Ventilation	Tolerates high CO_2 concentrations	One air exchange (3 h) every day	Two air exchanges (1 h) every 12 h	Two air exchanges (2 h) every 12 h	Three air exchanges (2 h) every 12 h
Time (days)	30–45	45–55	10–20	10	7–10

In contrast, the lowest *BE* obtained for the same strain inoculated in 2-kg substrate units was 22%. The highest *BE* values were obtained for the WC836 and CHFZ01 strains in 2-kg substrate units (30% and 32%, respectively). The determination of the average *BE* for each strain and each substrate unit was calculated as the average of the productive bags during each batch. The highest average *BE* was 31% for strain CHFZ01 in the 2 kg substrate units.

The production of *G. frondosa* at the pilot-scale level was performed in the Macromycetes Cultivation Unit at the Bioprocess and Agro-industry Plant. The cultivation rooms of this unit are 6 m^2 and 3 m high. These rooms were filled with 200 kg of inoculated substrate, as shown in Figure 46.

Table 30. Biological efficiency (*BE*) of four *Grifola frondosa* strains cultivated in 2 and 3 kg substrate units at the pilot scale

Strain	Higher *BE* (%)		Lower *BE* (%)		Average *BE* (%)	
	3 kg	2 kg	3 kg	2 kg	3 kg	2 kg
WC836	22.00±2.00	30.00±3.50	12.00±1.00	29.00±1.80	20.40±0.48	29.80±1.50
WC828	23.00±1.50	28.00±2.20	16.00±0.50	20.00±2.00	21.75±0.56	25.12±2.15
WC659	22.00±3.00	26.00±2.60	14.00±1.00	19.00±2.50	19.00±0.02	23.70±0.32
CHFZ01	26.00±2.50	32.00±1.00	20.25±1.50	28.00±0.70	23.13±1.01	31.00±1.21

Figure 46. Pilot scale *Grifola frondosa* cultivation in the Bioprocess and Agro-industry Plant at the Universidad de Caldas in Manizales (Colombia).

FINAL CONSIDERATIONS

Grifola frondosa is a mushroom that is not yet highly cultivated around the world. In tropical countries such as Colombia, there are no commercial cultures, and the knowledge level regarding this fungus is low. The Universidad de Caldas is positioned as the pioneer institution for engaging in research and development for the production of Maitake and related products in South America, from the implementation of its cultivation at different scales to the utilization of this mushroom for obtaining nutraceutical products and different bioactive metabolites.

In the Bioprocess and Agro-industry Plant at the Universidad de Caldas, a solid-state cultivation system for producing *G. frondosa* was developed at small and pilot scales on substrates based on oak sawdust supplemented with corn bran, sucrose, and calcium salts. The constraints for the growth and development of this mushroom on different types of lignocellulosic waste that is available in the coffee-growing region of Colombia were identified. In addition, the resulting fruiting bodies meet the quality parameters that were already established as suitable for direct consumption or transformation into a nutraceutical product.

References

Ainsworth, G. C. 2008. *Ainsworth & Bisby's Dictionary of the Fungi*. Wallingford, UK: CAB International.

Chang, S. T., and P. G. Miles. 2004. *Mushrooms. Cultivation, Nutritional Value, Medicinal Effect, and Enviromental Impact*. London: CRC Press.

Chen, A. W., P. Stamets, R. B. Cooper, N. Huang, and S. Han. 2000. "Ecology, morphology, and morphogenesis in nature of edible and medicinal mushroom *Grifola frondosa* (Dicks.: Fr.) S.F.Gray—maitake (Aphyllophoromycetideae)." *International Journal of Medicinal Mushrooms* 2 (3):221-28.

Chen, A. W., P. Stamets, N. L. Huang, and S. H. Han. 1999. "*Maitake at a glance.*" The Mushroom Company. Accessed October 2018. https://www.mushroomcompany.com/resources/maitake/ataglance.pdf.

Huang, N. L. 1997. "Japanese *Grifola frondosa*." In *Cultivation of Eighteen Precious and Delicious Edible Fungi*, edited by N. L. Huang, 145-53. Beijing: Chinese Agricultural Press.

Jong, S. C. 1978. "Conservation of the cultures." In *The Biology and Cultivation of Edible Mushrooms*, edited by S. T. Chang and W. A. Hayes, 119-36. New York: Academic Press.

Kirsop, B. E., and J. J. S. Snell. 1984. *Maintenance of Microorganisms: A Manual of Laboratory Methods*. Orlando, FL, USA: Academic Press.

Montoya, S. 2012. "*Obtención de enzimas lignocelulolíticas y polisacáridos a partir de residuos lignocelulósicos del Departamento de Caldas empleando macromicetos de pudrición blanca por fermentación sumergida y fermentación en estado sólido.*" PhD Diss., Facultad de Ciencias Agropecuarias, Universidad de Caldas. ["*Production of lignocellulolytic enzymes and polysaccharides from lignocellulosic waste from Department of Caldas employing white rot macromycetes by submerged fermentation and solid-state fermentation.*" PhD Diss., Faculty of Agricultural Sciences, Universidad de Caldas].

Montoya, S., O. Sánchez, and L. Levin. 2014. "Evaluación de actividades endoglucanasa, exoglucanasa, lacasa y lignina peroxidasa en diez hongos de pudrición blanca." *Biotecnología en el Sector Agropecuario y Agroindustrial* 12 (2):115-24. ["Evaluation of endoglucanase, exoglucanase, lacasse, and lignin peroxidase activities on ten white-rot fungi." *Biotechnology in the Agricultural and Agro-industrial Sector* 12(2):115-124].

Montoya, S., L. M. Varón, and L. Levin. 2008. "Effect of culture parameter on the production of the edible mushroom *Grifola frondosa* (maitake) in tropical weathers." *World Journal of Microbiology and Biotechnology* 24:1361-66.

O'Reilly, P. 2016. *Fascinated by Fungi: Exploring the Majesty and Mystery, Facts and Fantasy of the Quirkiest Kingdom on Earth*. Rhydlewis, Ceredigion, UK: First Nature.

San Antonio, J. P. 1979. "Stability of spawn stocks of the cultivated mushrooms stored for nine years in liquid nitrogen (-160°C to -196°C)." *Mushrooms Science* 10 (part 1):103-13.

San Antonio, J. P., and S. W. Hwang. 1970. "Liquid nitrogen preservation of the cultivated mushrooms, *Agaricus* (Lange) Sing." *Journal of the American Society for Horticultural Science* 95:565-69.

Shen, Q., and D. J. Royse. 2001. "Effect of nutrient supplements on biological efficiency, quality and crop cycle time of maitake (*Grifola frondosa*)." *Applied Microbiology and Biotechnology* 57 (1-2):74-78.

Shen, Q., and D. J. Royse. 2002. "Effects of genotypes of Maitake (*Grifola frondosa*) on biologycal activities, efficiency, quality and crop cycle." *Applied Microbiology and Biotechnology* 58:178-82.

Smith, D. 1993. "Culture collections." In *Genetics and Breeding of Edible Mushrooms*, edited by S. T. Chang, J. A. Buswell and P. G. Milles, 15-35. Philadelphia: Gordon & Breach Science.

Stamets, P. 1993. *Growing Gourmet and Medicinal Mushrooms*. Hong Kong: Ten Speed Press and Mycomedia.

Stamets, P. 2000. *Growing Gourmet and Medicinal Mushrooms*. Berkeley: Ten Speed Press.

Wu, J. L., J. C. Lin, M. C. Fu, and S. T. Wei. 1997. "*Grifola frondosa*." In *Cultivation of Eighteen Precious and Delicious Edible Mushrooms*, edited by N. L. Huang, 130-45. Beijing: Chinese Agricultural Press.

Chapter 6

COMMERCIAL SCALE CULTIVATION OF *GRIFOLA FRONDOSA*

INTRODUCTION

The artificial cultivation of *Grifola frondosa* in plastic bags began in Japan in the 1980s, which led to a niche market in that country and in the United States. Subsequently, the demand has extended to some countries in Europe and to Australia. As noted in the previous chapter, factors that have constrained Maitake cultivation include the strict control of its environmental conditions and the availability of suitable raw materials for its commercial production.

The industrial production of Maitake with commercial purposes is performed on synthetic substrates contained in polypropylene bags in countries where this mushroom is cultivated. Oak sawdust is a common substrate that is used for the commercial production of *G. frondosa* in the United States and Japan (Lee 1994, Royse 1997), with the use of beech and larch sawdust in Japan being less common (Kirchhoff 1996, Stamets 2000). In China, husks and cottonseed waste are used as a substitute for wood sawdust with acceptable yields (Zhao, Yang, and Li 1983). Bran of cereals, such as rice bran, oat bran, corn bran, and wheat bran, among others, are used as nutritional supplements for Maitake cultivation, as well as whole cereal grains (e.g., red sorghum), corn flour, and soybean cake (Shen and Royse 2002a).

Information on *G. frondosa* cultivation techniques is still limited, even more so in terms of commercial-scale methods of cultivation. Currently, in many countries, the cultivation techniques used for other macromycetes species continue to be adapted to cultivate the Maitake. For example, techniques for cultivating *Lentinus edodes* (Shiitake) have been adopted for *G. frondosa* production because both species are white rot fungi and because the Shiitake mushroom has been widely studied, thus resulting in the development of efficient cultivation methods, genotypes, and effective nutritional formulations. The fast growth of Maitake production has forced the development of more efficient substrate formulations to improve the yield and quality of its fruiting bodies and shorten its cultivation cycle. Shen and Royse (2002a) created an experimental design to determine the effects of selected nutrient supplements and their levels on the culture cycle time, biological efficiency (*BE*), yield and quality for the most widely used commercial isolate of *G. frondosa* (WC828) in the United States. These authors found significant differences between the different formulations. They also identified combinations of nutritional supplements among those evaluated. However, greater efforts are needed for the commercial implementation of Maitake culture to improve the *BE* and the quality of the fruiting bodies and to shorten the culture cycle, among other aspects. The expansion of commercial *G. frondosa* cultivation in other countries, especially in countries with tropical weather, is still in progress.

In this chapter, a model for the commercial cultivation of *Grifola frondosa* in a rural zone of Villamaría, a town located in a mountainous region of Colombia, which is a tropical country, is presented through several research cases. This information corresponds to the results obtained by the authors of this book during the execution of the research project entitled "Development and evaluation of the parameters of semi-industrial scale cultivation of the fungus *Grifola frondosa* in tropical climates and its transformation for commercial purposes as nutraceutical products" from the Call 523-2011 framework "Project Bank for Creation of Technology-based

Companies or Business Units" as funded by the Colombian Administrative Department of Science, Technology and Innovation (Colciencias).

DEVELOPMENT OF *G. FRONDOSA* CULTURE AT THE INDUSTRIAL SCALE

In Latin America countries and in tropical countries in general, the commercial-scale cultivation of *G. frondosa* has not been reported. The probable reasons why Maitake cultivation is not so widespread in these areas and in the world in general are varied. The raw materials required by this mushroom are very specific, with low availability and high cost. In addition, the environmental conditions that must be provided to the culture for its development at different stages require the implementation of an infrastructure that generates high costs. These issues are related to the needs of the *G. frondosa* species and do not allow for the implementation of alternatives that would make the artificial cultures of Maitake efficient. Despite these challenges, the protocols for cultivating this mushroom at small, pilot, and industrial scales have been developed in the Bioprocess and Agro-Industry Plant at the Universidad de Caldas as a result of different research projects and initiatives intended for the implementation of Maitake production under the conditions of a tropical country, such as Colombia, and by considering the commercialization opportunities for Maitake-derived products that have arisen in recent years.

Infrastructure Required for Commercial Maitake Production

The cultivation of mushrooms requires a controlled environment to obtain high-quality fruiting bodies and for the production processes to be stable and exhibit competitive biological efficiencies, and the cultivation of *G. frondosa* mushrooms is no exception. In the previous chapter, the protocols for producing *G. frondosa* at small and pilot scales were described.

The experience gained during the development of these protocols provided a basis for implementing Maitake production by solid-state cultivation at the industrial scale. For this purpose, it was necessary to define the infrastructure required to produce Maitake fruiting bodies. The following research case illustrates this topic.

Research Case 20. Preparation of the Infrastructure for Producing G. frondosa at the Industrial Scale

To perform cultivation tests for the industrial-scale production of Maitake mushrooms, cultivation facilities that were previously employed for the production of button mushroom (*Agaricus bisporus*) were adapted for *G. frondosa* production. These facilities are located in La Vega farm, Lot 2 at Gallinazo vereda (rural district), km 1 towards the aqueduct of the Municipality of Villamaría, Department of Caldas, Colombia, which has an average temperature of 16-18°C and a height of 2,250 meters above sea level. It is worth highlighting that the production of Maitake mushrooms at the industrial scale requires an average temperature of 16-18°C for fruiting; therefore, this temperature was possible to ensure at La Vega farm, Lot 2. Similarly, the cultivation rooms formerly used for growing button mushrooms fulfilled the requirements for adaptation into cultivation rooms for producing other macromycetes species at the industrial or semi-industrial scale. The dimensions of these cultivation rooms are 8 m×10 m×4 m (width×depth×height). Two external pictures of the cultivation rooms during the adaptation renovations are shown in Figure 47.

Three cultivation rooms were divided in half lengthwise with a Super board (fiber cement panels) wall covered with plastic to obtain six smaller cultivation rooms of 40 m^2 each. The floor of the cultivation rooms was rebuilt with a concrete plate to eliminate the existing pores and damage, the roof leaks were repaired, the ceilings were painted internally, an individual door was provided for each room to individualize each environment, shelves were built with bamboo and mats were placed for the distribution of substrates in each room (see Figure 48). In the same way, environmental control was installed in each cultivation room to control the relative humidity (10-100%), temperature, air exchange, and light intensity as shown in Figure 49. The environment inside the cultivation rooms is controlled through a PLC, which is connected to a water sprinkler to maintain the necessary relative humidity and to heaters to control the temperature during the incubation phase of the substrate blocks.

After the adaptation and conditioning of the old facilities for button mushroom production, functional and suitable cultivation rooms for *G. frondosa* production were operated to perform the solid-state cultivation of Maitake mushrooms under industrial conditions.

(a) (b)

Figure 47. Cultivation rooms at La Vega farm Lot 2 during the adaptation renovations for cultivating *Grifola frondosa*. (a) Adaptation of the external part of the cultivation rooms: Windows for air exchange, doors, and painting. (b) Roof with curved tiles for the cultivation rooms.

Cultivation Conditions for Maitake Production at the Industrial Scale

The operating and environmental conditions for *G. frondosa* production at the industrial scale do not necessarily coincide with the conditions applied to the small or pilot-scale production of this mushroom. Therefore, once the required infrastructure is set up, it is necessary to perform several tests in the industrial facilities to define the modifications needed in the operating parameters for the production and startup of the cultivation process at larger scales.

(a) (b) (c)

Figure 48. Details of civil adaptation in the interior of the cultivation rooms. (a) Details of individual doors for each cultivation room with axial 14-inch fans. (b) Details of the division panel covered with plastic in a cultivation room with bamboo shelves and matting and whitewashing for the disinfection of the rooms and the correction of floor pores. (c) View of a cultivation room with the door and fan.

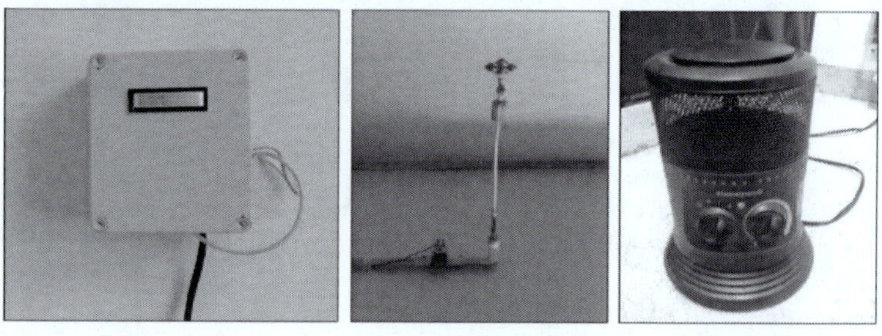

(a) (b) (c)

Figure 49. Accessories for environment control in cultivation rooms: (a) PLC for programming the devices that control the humidity, air exchange, and room temperature; (b) water sprinkler; and (c) heating device for the temperature control of the cultivation rooms.

To implement the large-scale cultivation of *G. frondosa*, it is necessary to know the requirements of the species and to define the environmental conditions required relative to the conditions that were set for cultivation at lower production scales. In the following research case, these conditions were defined for Maitake production in the cultivation rooms described in Research Case 20.

Research Case 21. Definition of Cultivation Conditions for the Production of G. frondosa at the Industrial Scale and the Production of Commercial Spawn

Modification of Cultivation Conditions

The operating and environmental conditions for cultivating *G. frondosa* mushrooms at the commercial scale were defined and established based on the cultivation conditions for this same mushroom at the pilot-scale level, which were discussed in the previous chapter (see Research Cases 18 and 19). Several trials were performed on four *G. frondosa* strains (WC836, WC828, and WC659 from the Pennsylvania State University Mushroom Culture Collection, United Stated and CHFZ01 from the Fujian Agriculture and Forestry University, China). After these trials, the following modifications were established to perform the process at the industrial scale:

- During the formulation of the substrates, calcium sulfate pentahydrate was introduced, which provides texture to the culture medium. Thus, the formulation yielding the best biological efficiency (*BE*) for the commercial scale is (on a dry basis) as follows: 74% oak sawdust, 23% corn bran, 1% $CaCO_3$, 1% $CaSO_4$, and 1% sucrose. The moisture of the substrates was adjusted to 58%.
- The spawn for substrate inoculation for fruiting body production was prepared on a mixture of agroindustrial wastes (see below).
- The substrate unit was increased from 1 kg to 2 kg and packed in bioriented polypropylene bags (this size was defined during pilot-scale experimentation).
- For the industrial-scale culture, no thermal shock was performed at 10°C as indicated for the small-scale cultivation process. It was possible to obtain acceptable productivities with three of the four strains evaluated at 2,250 meters above sea level and an average environmental temperature of 17°C when lowering the fructification

temperature down to 15°C for 3 or 4 days by manipulating the environmental conditions of the cultivation rooms.

- A single harvest was obtained from each production batch since tests performed on the pilot scale showed that the fruiting bodies of the second harvest were very small and had low weights.

Production of Third-Generation Maitake Spawn from Lignocellulosic Waste

Figure 50. Progress of *Grifola frondosa* colonization during spawn production from lignocellulosic waste: (a) 15 incubation days, (b) 25 incubation days, (c) 35 incubation days and (d) 45 incubation days.

The production of commercial spawn (third-generation spawn) of *G. frondosa* was performed on a mixture of lignocellulosic waste with the following composition (on dry basis): 74% oak sawdust, 23% corn bran, 1% calcium sulfate, 1% calcium carbonate, and 1% sucrose. The mixture moisture was adjusted to 60%. This modification was made based on the research developed in a previous study (Montoya, Varón, and Levin 2008). In that research, a reduction in the adaptation stage at the incubation phase was achieved by considering the fact that the tested *G. frondosa* strains were adapted to the culture medium defined for the production of carpophores in a preliminary batch that was prepared from 200 kg substrate for fructification purposes. In addition, the incubation time was reduced by over 9 days in the colonization phase since the time for complete colonization reached 21 days. Figure 50 shows the progress of the substrate colonization by the Maitake mushroom mycelium when third-generation spawn is produced.

Production Cycle of *G. frondosa* at the Industrial Scale

The cycle used for producing Maitake mushrooms at the industrial scale involves the preparation of the substrate, its inoculation by using the third-generation spawn, an incubation phase (including the first subphase in which the mushroom adapts to the substrate and the subsequent complete colonization by the mycelium and the second subphase in which the mycelial coat is formed), the phase of primordia formation, and the development of fruiting bodies. In the following research case, this production cycle for processing 7 tonnes of substrate is briefly discussed.

Research Case 22. Maitake Production Cycle at a Commercial Scale

To produce *G. frondosa* on an industrial scale, the substrate formulation was as follows (on dry basis): 74% oak sawdust, 23% corn bran, 1% calcium sulfate, 1% calcium carbonate, and 1% sucrose. The moisture of the substrate mixtures was adjusted to 58%.

Seven 1,000-kg substrate batches were packed into bioriented polypropylene bags containing 2 kg of substrate each. Each bag had a cotton plug that had been previously sterilized and that was secured with a PVC ring to allow respiration of the biomass. The substrate packed in bags was subjected to sterilization at 121°C for 2 h with subsequent cooling to room temperature. The substrate was inoculated with the spawn that had previously developed on the same substrate formulation. The concentration of the added production inoculum was 5% (w/w) on a wet substrate basis.

The stages for substrate preparation were performed in the Bioprocess and Agro-industry Plant at the Universidad de Caldas, which is located near the industrial facilities at La Vega farm, Lot 2 (15 min away by road). In particular, substrate sterilization was performed in a horizontal autoclave with a processing capacity of 500 kg of solid substrate (see Figure 51).

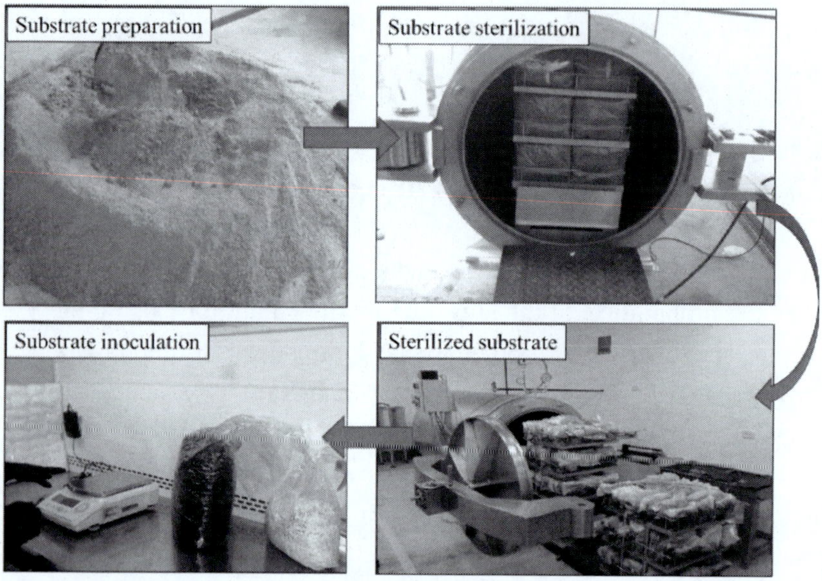

Figure 51. Overall stages for preparing substrates at a 1-ton scale for *Grifola frondosa* production at the Bioprocess and Agro-industry Plant.

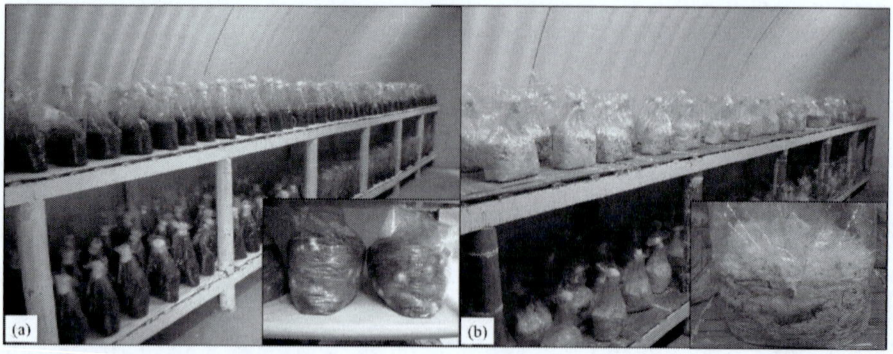

Figure 52. Mycelium development of *Grifola frondosa* on the substrate in the cultivation rooms. (a) Substrate units 15 days after inoculation. (b) Substrate units 75 days after inoculation.

After the inoculation of the substrates, they were incubated in cultivation rooms at 25°C for 70-90 days at the commercial-scale facilities until the formation of the mycelial coat with orange pigmentation and exudates of the same color. The incubation phase consisted of two subphases. The first one corresponded to the adaptation stage and the complete colonization of the mycelium on the substrate; for this purpose, the temperature was maintained at 25°C, and the relative humidity was 50-60% during the first 30 days, which are decisive factors in this process since Maitake mushrooms grow quite slowly and are sensitive to competitors.

Next, the second incubation subphase was started, in which the formation of the mycelial coat and the pigmentation began; the exudates that formed were clearly visualized in the substrate units. The average time of this subphase was 30-40 days, and the substrate units were maintained at 20°C and 65% relative humidity. After this subphase, the substrates were ready for primordia induction. In Figure 52, the progress of the substrate during the incubation phase is clearly observed (notice the change in the substrate pigmentation).

To induce the formation of Maitake fruiting bodies, it is necessary to perform a light shock between 100 and 150 luxes for 3 to 4 days and to decrease the temperature to 15-16°C; then, the temperature may be maintained at 18°C.

The primordia should be formed during the week that follows. Figure 53 shows how the beginning of Maitake fruiting body formation is identified, demonstrating a black-grayish exudate located on the substrate. To obtain the mushrooms at the point of harvest, they must undergo well-differentiated changes, which are shown in Figure 54. These states are briefly described as follows:

- Brain state. On the surface of the substrate and after the black-gray pigmented exudate forms, a brain-shaped mass of the same dark color is formed, as shown in Figure 54 (a and b).

- Cauliflower state. The growth and development of the mushroom with the elongation of the amorphous bodies and cap definition. Two previous substrates are presented, corresponding to the forms following the brain state (first, the conformation and then the definition of the caps, and second, their widening and attenuated coloring), as shown in Figure 54 (c and d).
- Mature mushroom state. When ready for harvest, complete elongation of the caps, discoloration of the body until light tan brown and the filling of the amorphous bodies with ramifications, as well as the complete formation of the spore system, are observed. These spores are initially white, and the mushroom presents a coral appearance with a pleasant aroma of fresh nuts. The mushroom in its mature state is shown at the point of harvest in Figure 54 (e and f).

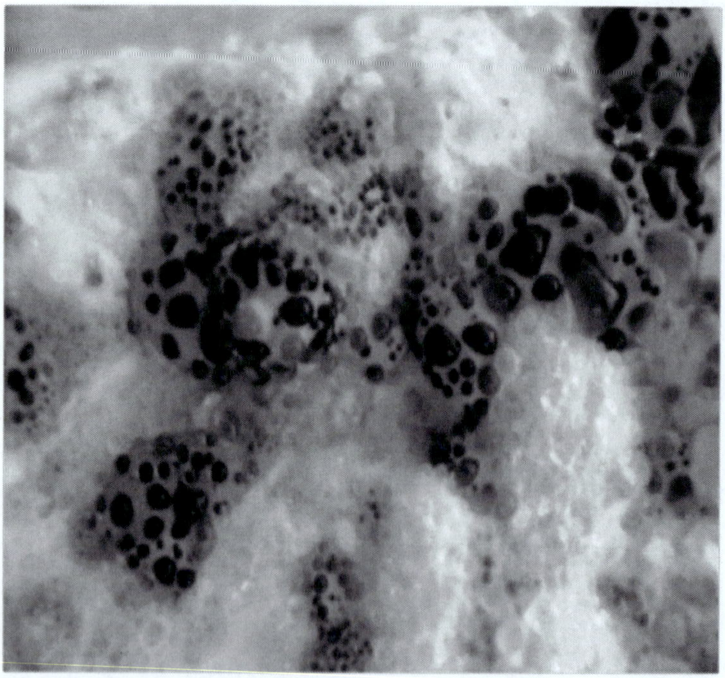

Figure 53. Pigmented exudate is grayish black on a sector of the substrate as an indicator that the *Grifola frondosa* fruiting body has formed.

The main growth and development stages during G. frondosa cultivation on an industrial scale are shown in Figure 55. In general, the process begins with the loading of 1 tonne of bagged substrate that had been inoculated in cultivation rooms measuring 40 m2 and 4 m high, which are suitable for the commercial production of fruiting bodies, and it ends with the harvesting of the Maitake fruiting bodies.

Table 31 lists all the environmental changes according to the phases that should be taken into account during Maitake cultivation at the industrial scale after substrate inoculation. Notice the increase in the duration of the air exchange for primordia formation and fructification phases compared to the values indicated in Table 29 (see previous chapter), as well as the change in the process time of each phase.

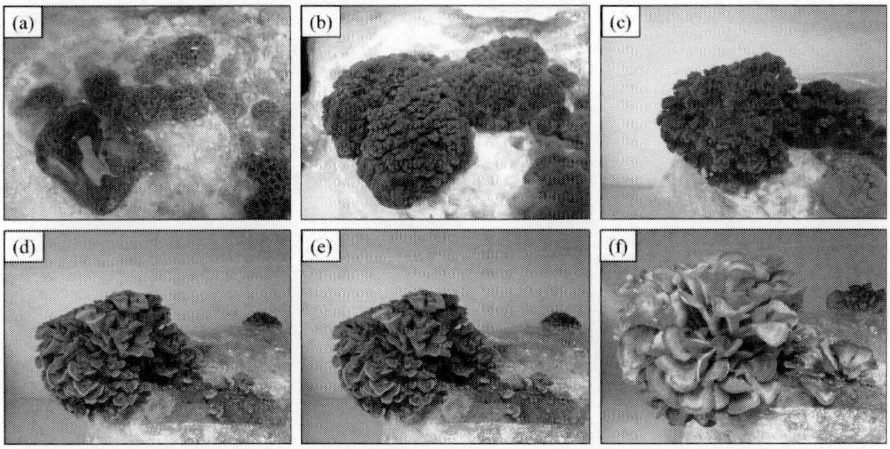

Figure 54. Growth stages of *Grifola frondosa* mushrooms: (a) Formation of primordia with dark gray pigmentation; (b) primordium formed in the shape of a brain; (c) a mushroom with the form of cauliflower; (d)-(e) larger posterior mushroom with a cauliflower shape; and (f) a mushroom that is ready for harvest.

Figure 55. Main stages of *G. frondosa* growth and development at the industrial scale in cultivation rooms: (a) Substrate units in the first incubation subphase incubation; (b) substrate units in the second incubation subphase; and (c) substrate units in the production phase.

Harvest and Post-Harvest Treatment of Maitake Mushrooms

After the fruiting bodies of *G. frondosa* are harvested, the yields should be calculated in terms of biological efficiency to assess the process performance. These carpophores must be properly packed so as not to reduce their quality and to meet market requirements.

Quality assessments may be performed using the quality guidance for Maitake established by Shen and Royse (2002b). These topics are discussed in the following research case.

Table 31. Summary of environmental conditions required to produce Maitake mushrooms on synthetic substrate at an industrial scale

Parameter	Incubation		Primordia formation		Fruiting development
	Subphase I	Subphase II	Induction	Formation	
Temperature (°C)	20–25	20	15–16	16–18	16–18
Relative humidity (%)	50–60	60	60	70–75	80–85
Light intensity (luxes)	Not required	Not required	50–150	100–150	100–150
Ventilation	Tolerates high CO_2 concentra-tions	One air exchange (3 h) every day	Two air exchanges (2 h) every 12 h	Two air exchanges (3 h) every 12 h	Three air exchanges (3 h) every 12 h
Time (days)	30–45	30–40	3–4	7	7–10

Research Case 23. Yields and Quality of Maitake Fruiting Bodies

For carpophores produced at a commercial scale (see Research case 22) by employing the four *G. frondosa* strains studied here, the corresponding *BE* values were calculated according to equation (1). For these strains, the higher, lower, and average biological efficiencies obtained in the industrial facilities located in La Vega farm, Lot 2 are presented in Table 32. It is highlighted that strains WC836 and WC659 were the ones that showed the highest and lowest *BE*s, respectively, of all the evaluated substrates. The *BE*s were calculated for a single harvest in all cases. In addition, the highest average *BE* was found for strain CHFZ01.

Table 32. Biological efficiency (*BE*) of four *G. frondosa* strains cultivated in 2-kg substrate units at a commercial scale.

Strain	Higher *BE* (%)	Lower *BE* (%)	Average *BE* (%)
WC836	34.00±1.50	16.10±1.20	23.20±2.20
WC828	30.00±2.00	20.00±1.70	23.81±2.40
WC659	28.00±1.00	14.00±2.00	20.56±2.10
CHFZ01	30.75±2.30	20.25±2.10	24.70±1.40

The attained *BE*s were influenced by the components in the substrate, the environmental conditions throughout the culture cycle, and the production scale. These results compare favorably with the data obtained by other researchers. The values for the biological efficiencies of Maitake mushrooms have been reported over several varying ranges: 7.5%-15.5% (Kirchhoff 1996), 10%-29% (Stamets 1993), 14% (Chalmers 1994), 16%-27% (Royse 1997), 25%-32.5% (Mayuzumi and Mizuno 1997), and 35.8%-39.5% (Shen and Royse 2001). The *BE* values reached in this research were only below the highest interval of the above-cited reports, which corresponds to the work of Shen and Royse (2001). However, it is worth emphasizing that all the reported *BE* values were obtained at a small scale and not on an industrial scale, as with the values presented here.

The harvested mature fruiting bodies are observed in Figure 56. The overmature carpophores are lighter in color and have very developed caps, diffuse stipes, and high spore formation. In contrast, mature fruiting bodies may be described as those that are not fully elongated and for which the stipe has not lost its shape, sporulation has barely begun, and the coloration is still brown on the upper part of the caps and whitens on anteriorly.

Figure 56. Maitake carpophore harvest.

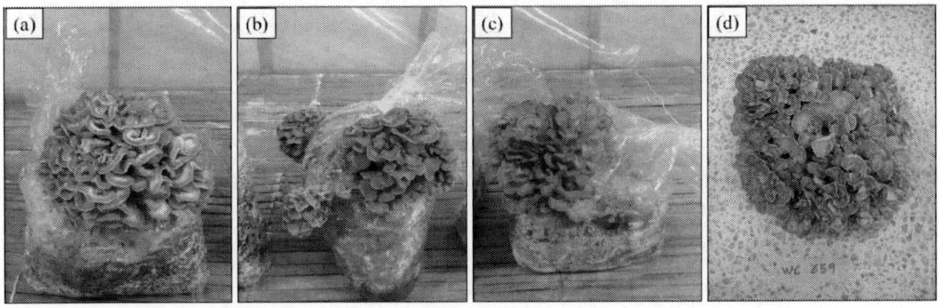

Figure 57. Fruiting bodies of *G. frondosa* strains grown on an industrial scale: (a) Strain CHFZ01; (b) strain WC828; (c) strain WC836; and (d) strain WC659.

High-quality Maitake fruiting bodies produced on substrates that are formulated with mixtures of oak sawdust, corn bran, and calcium salts can be obtained on an industrial scale. In this research, quality 1 carpophores were obtained according to the quality scale defined by Shen and Royse (2002b), i.e., the highest quality for Maitake fruiting bodies (dark gray basidiomes with well-formed bodies and a firm texture). However, considering this quality scale, strain WC659 was the one digressed from the quality standards developed by these authors, as the corresponding fruiting bodies had brown tones but strong pilei, although they were narrower and smaller than those of the other three strains (see Figure 57). Because the carpophores of this strain had a pleasant appearance and aroma, they were classified as quality 1. This scale of quality is of great importance for the sale of fresh fruiting bodies and for the synthesis of bioactive substances from each carpophore. For this reason, the determination of the quality of these fresh basidiomes was performed in this study.

The Maitake fruiting bodies obtained from the four strains were harvested at the appropriate point of development. During the post-harvest period, Maitake mushrooms may be packed in plastic trays coated with food packaging film and stored under refrigeration at 3-5°C for approximately one week to keep them fresh and suitable for consumption as a fresh food. Alternatively, they can be subjected to minimum processing by dehydration and grinding for sale as a nutraceutical product. Thus, the mushrooms produced here were taken to a minimum-processing room (located at Bioprocess and Agro-industry Plant), within which they were cut into pieces for drying. Then, a forced-convection tray dryer was loaded with 250 g of fresh mushrooms that were subjected to a drying process at 40°C for approximately 24 h. The mushrooms were dehydrated at this low temperature to minimize the thermal damage that could occur in the bioactive substances contained in the fruiting bodies. The final moisture of the dried mushrooms was between 5% and 7%. Subsequently, the mushrooms were ground to obtain a fine powder, which was packed in 50-g jars for sale or as feedstock for extracting biologically active substances. This procedure will be described in Chapter 9.

FINAL CONSIDERATIONS

To cultivate *G. frondosa* mushrooms on an industrial scale, several changes should be made in the process to adapt the small-scale processing conditions to the specific needs of commercial production. These modifications are primarily related to the substrate formulation, size of the substrate units, duration of the different phases of Maitake development, and the conditions of the thermal shock in relation to the ambient temperature in tropical weather. In particular, it is recommended that growers obtain only a single harvest from each substrate unit since the fruiting bodies from the second harvest are small and have low weights. The four *G. frondosa* strains evaluated in the research cases from this chapter are suitable for producing fruiting bodies both for their commercialization as a fresh food and for their transformation into Maitake-based dietary supplements or valuable metabolites.

REFERENCES

Chalmers, W. 1994. "*Specialty mushrooms: cultivation of the maitake mushroom.*" Mushroom World June:62.

Kirchhoff, B. 1996. "*Investigations of genotypes and substrates for the fruitbody production of Grifola frondosa* (Dicks.:Fr.)." Mushroom Biology and Mushroom Products: Proceedings of the 2nd International Conference, University Park, PA, USA.

Lee, E. 1994. "Production of shiitake, oyster and maitake mushrooms in Connecticut." *The Mushroom Growers´ Newsletter* 3:11-14.

Mayuzumi, Y., and T. Mizuno. 1997. "Cultivation methods of maitake (*Grifola frondosa*)." *Food Reviews International* 13:357-364.

Montoya, S., L. M. Varón, and L. Levin. 2008. "Effect of culture parameter on the production of the edible mushroom *Grifola frondosa* (maitake) in tropical weathers." *World Journal of Microbiology and Biotechnology* 24:1361-1366.

Royse, D. J. 1997. "Specialty mushrooms and their cultivation." *The Horticulture Journal* 19:59-97.

Shen, Q., and D. J. Royse. 2001. "Effect of nutrient supplements on biological efficiency, quality and crop cycle time of maitake (*Grifola frondosa*)." *Applied Microbiology and Biotechnology* 57 (1-2):74-78.

Shen, Q., and D. J. Royse. 2002a. "Crop cycle time, yield and quality of maitake *Grifola frondosa* as influenced by nutrient supplements." *4th International Conference for Mushroom Biology and Mushroom Products* (The 4th ICMBMP), Cuernavaca, Mexico.

Shen, Q., and D. J. Royse. 2002b. "Effects of genotypes of Maitake (*Grifola frondosa*) on biologycal activities, efficiency, quality and crop cycle." *Applied Microbiology and Biotechnology* 58:178-182.

Stamets, P. 1993. *Growing Gourmet and Medicinal Mushrooms*. Hong Kong: Ten Speed Press and Mycomedia.

Stamets, P. 2000. *Growing Gourmet and Medicinal Mushrooms*. Berkeley: Ten Speed Press.

Zhao, Z. G., X. L. Yang, and Z. M. Li. 1983. "The cultivation of *Polyporus frondosa*." *Edible Fungi* 5:8-9.

Chapter 7

RESEARCH TRENDS ON *GRIFOLA FRONDOSA* CULTIVATION AND ITS BIOLOGICALLY ACTIVE COMPOUNDS

INTRODUCTION

The development of solid-state cultivation methods for *G. frondosa* under controlled conditions is much more recent than that of cultures of other macromycetes with culinary and medicinal attributes. Several types of solid cultures for producing fruiting bodies have been developed empirically by adapting research methodologies and simulating wild substrates and environmental conditions that correspond to the places where Maitake basidiomes have been found. For this reason and because of the specific nature of Maitake culture, the generation of new alternatives for industrial-scale cultivation has not been simple. To this end, *G. frondosa* production has been developed in bags containing 1-5 kg of substrate and with many constraints regarding the selection of feedstocks used to prepare the substrates. In addition, this culture requires specific environmental parameters for its growth and development, which has generated limitations in terms of the geographical locations of the cultures. Moreover, *G. frondosa* is not a completely saprophytic fungus because it grows as a parasite on the base of living trees, both coniferous and deciduous (especially species of the genera *Quercus*, *Acer*, *Betula*, *Carpinus*, *Castanea*, *Fagus*, and *Ulmus*), in the temperate forests of Asia, Europe, and North America (Carlile, Watkinson, and Gooday 2001, Gilbertson and Ryvarden 1987, O'Reilly 2016). This challenge has made its artificial cultivation even more difficult. Therefore, different research efforts are being undertaken throughout the world to improve Maitake cultivation by assessing alternative mixtures of feedstocks (primarily waste materials) as well as reducing production costs by increasing the efficiency of the solid-state cultivation itself.

Notably, the important spectrum of biologically active compounds that may be obtained from the mycelium and fruiting bodies of *G. frondosa* is another ongoing research direction in many universities and research centers around the world. In particular, the extraction, characterization, and determination of the pharmacological properties of the polysaccharides from Maitake mushrooms are of paramount importance for developing nutraceutical products and active compounds that help to treat several diseases, such as cancer.

This chapter covers some of the main research trends related to the potential use of alternative feedstocks for *G. frondosa* culture, the modeling of solid-state Maitake cultivation to enhance our knowledge of the process, and the extraction and characterization of biologically active compounds from this mushroom.

G. frondosa Culture Using Alternative Feedstocks

One of the main challenges in cultivating *G. frondosa* is the selection of the feedstocks. There is not enough information about the potential raw materials for Maitake culture when considering that its production is relatively recent in the world. To reduce production costs, these materials should be available and inexpensive. Agricultural and agro-industrial waste are employed as feedstocks for most mushrooms. However, the range of traditional wastes and residues from which the Maitake can be produced is reduced. For this reason, it is necessary to engage in research to assess alternative waste materials by accounting for their availability in the countries where Maitake production is already established or where the conditions for its production are favorable (as in some tropical countries) and there is a potential market for its commercialization. In the following subsections, two research and development directions for Maitake production from alternative feedstocks are disclosed.

Solid-State Cultivation of Maitake by Using Grasses

Before the 1980s, logs and sawdust were the only materials used for *G. frondosa* cultivation, similar to the cultivation of *G. lucidum*. Considering the need for a sustainable Maitake production that does not involve the use of forestry resources, the solid-state bag cultivation of this mushroom by using substrates based on wild grasses (JUNCAO, see Chapter 3) was successfully achieved. The development conducted by Chinese researchers at the Fujian Agriculture and Forestry University at Fuzhou (China) for solid-state *G. frondosa* cultivation by applying JUNCAO technology comprises the following process steps (Lin and Lin 2014):

- JUNCAO pretreatment. This step begins with the cultivation of grass species suitable for cultivating Maitake mushrooms, with the subsequent drying and grinding of the grass to conserve it.
- Formulation of alternative substrates using different grass species: these formulations are detailed in Table 33.
- Preparation of substrate units in plastic bags containing 0.5, 0.8 and 0.9 kg of substrate and plastic bottles containing 0.25 kg of substrate.
- Substrate sterilization at 121°C and 15 psig for 2 h or 100°C and atmospheric pressure for 10 h.

Table 33. Substrate formulations containing different grass species (JUNCAO) for Maitake cultivation

Material	F1	F2	F3	F4	F5	F6
Miscanthus floridulus	61%	55%	-	25%	-	-
Pennisetum purpureum	-	20%	60%	20%	33%	78%
Dicranopteris dichotoma	15%	-	17%	-	-	-
Sorgum propinquum	-	-	-	30%	-	-
Corn cobs	-	-	-	-	53%	-
Wheat bran	15%	18%	15%	18%	12%	10%
Rice bran	7%	5%	5%	5%	-	10%
Gypsum powder	2%	2%	3%	2%	2%	2%

Water content: 58-62%; pH value: 6.0-7.0.
Source: Lin and Lin (2014).

- Inoculation in a close, dry, clean, and disinfected room with relative humidity below 60%; the substrate temperature should be less than 30°C.
- Incubation of inoculated substrate units at 24-28°C and relative humidity lower than 70%.
- Primordia induction at 18-23°C and 50-150 luxes.
- Growth and development of fruiting bodies at 16-18°C, 200-500 luxes and fresh air in excess.
- The harvest is performed when the tubules (spore-forming pores) have been completely formed in the carpophores.

Every 100 g of dried Maitake mushroom contains 19.7 g of crude protein, 3.2 g of fat, 61.1 g of total sugar, 9.74 g of fiber, and 6.3 g of ash (Lin and Lin 2014). The JUNCAO technology for producing *G. frondosa* was transferred to Japan by Fujian Agriculture and Forestry University in 1992.

Utilization of Waste from the Coffee Agro-Industry

The possibility of producing *G frondosa* from oak sawdust or shavings in countries with tropical weather was discussed in Chapters 5 and 6 for Colombia. However, it is important to conduct research on other solid waste materials available in such countries that could be employed for substrate formulation. The following research case illustrates an attempt to use other nontraditional residual materials for cultivating *G. frondosa* in Caldas, a coffee-growing region of Colombia.

Research Case 24. Evaluation of Alternative Feedstocks for G. frondosa Production

One of the purposes of this research case was to evaluate the growth and development of the Maitake on substrates formulated from materials different from those that the mushroom uses for its growth in the wild. For this purpose, a strain of the mushroom *G. frondosa* (PSUMCC 922) obtained from the Pennsylvania State University Mushroom Culture Collection (USA) was employed to evaluate two different substrate formulations based on Colombian agro-industrial waste. One of these formulations included coffee spent-ground, an agro-industrial waste typical of the region of Caldas, which is one of the first departments (equivalent to provinces) to produce coffee in Colombia and has several coffee-processing factories. The outcomes presented in this research case are the results of two previous studies (Montoya, Orrego, and Levin 2012, Montoya, Varón, and Levin 2008).

The first formulation had the following composition (on a dry basis): 75% oak sawdust, 23% corn bran, 1% sucrose, and 1% calcium carbonate. In the second formulation, all of the oak sawdust content was replaced by 50% oak sawdust and 25% coffee spent-ground; the remaining components had the same concentrations as the first formulation. The rest of the procedure was described in Research Case 18 (see Chapter 5).

One of the most striking results of this research was that the substrate containing coffee spent-ground did not produce *G. frondosa* fruiting bodies. The spent substrates (the substrate blocks remaining after the cultivation process) were subjected to a complete bromatological analysis (see Table 34).

From this analysis, it was determined that the zinc content in the spent substrate corresponding to the formulation that included coffee spent-ground (290 ppm) was significantly higher than the zinc content in the spent substrate corresponding to the first formulation without the coffee grounds (17 ppm). The results presented for the bromatological analysis corresponded to the average value of 20 cultivation batches for each substrate formulation. The substrates that fructified were those with formulations based on oak sawdust, with an average biological efficiency of 35.3% (see equation 1).

Table 34. Bromatological analysis of the spent substrate from two formulations used for Maitake production (% on dry basis)

Components	Oak sawdust-based substrate	Substrate based on oak sawdust and coffee spent-ground
Moisture (%)	52.79 ± 0.6	52.87 ± 0.86
Dry matter (%)	45.21 ± 0.6	47.13 ± 0.86
Total nitrogen (%)	0.23 ± 0.01	0.70 ± 0.011
Total protein (%)	1.44 ± 0.012	4.37 ± 0.013
Total lipids (%)	0.45 ± 0.013	1.18 ± 0.013
Crude fiber (%)	52.21 ± 1.25	38.69 ± 1.54
Total ash (%)	2.40 ± 0.37	4.05 ± 0.41
P (%)	0.14 ± 0.005	0.03 ± 0.001
Ca (%)	0.40 ± 0.003	0.37 ± 0.002
Mg (%)	0.10 ± 0.001	0.03 ± 0.0015
K (%)	0.28 ± 0.006	0.14 ± 0.002
Na (%)	0.01 ± 0.001	0.02 ± 0.001
Fe (ppm)	145.47 ± 3.87	136.72 ± 1.275
Mn (ppm)	29.90 ± 2.134	26.25 ± 1.356
Zn (ppm)	17.30 ± 1.456	290.25 ± 5.467
Cu (ppm)	2.10 ± 0.664	3.17 ± 0.234

Source: Montoya, Varón, and Levin (2008).

From the data in Table 34, a significant difference in the nitrogen content of the two substrate formulations is observed as well, with 0.23 and 0.70 for oak sawdust-based substrate and for coffee spent-ground-containing substrate, respectively. This difference suggests that the coffee residue increased the nitrogen contents of the substrates.

At this point, it is important to note that the bromatological analyses were performed after the fruiting bodies were harvested from the substrates formulated with oak sawdust; the substrates containing coffee spent-ground were analyzed at the same time as the spent substrates containing oak but that never produced carpophores. It is important to note that high nitrogen contents are not suitable for Maitake fructification. Similarly, the lipid and ash contents were higher in the spent substrate containing coffee spent-ground compared to the spent substrate without this component. Thus, the ash content present in the substrates with coffee spent-ground may interfere with the formation of Maitake carpophores due to their high mineral levels (especially metals).

To find if high zinc contents in the substrate inhibit or limit the growth rate of *G. frondosa*, Petri dishes (90 mm diameter) containing PDA medium supplemented with 25, 100, 250 or 500 ppm $ZnCl_2$ were inoculated and then incubated at 25°C for 21 days. The radial growth was measured in two perpendicular directions from the edge of the inoculum to the advancing margin of the colony. The radial growth rates (in mm per day) were determined.

The growth rates on PDA medium after 21 days were slightly affected by zinc concentrations of up to 100 ppm. The growth inhibition was higher, and longer lag-phases during the cultivation were found for 250–500 ppm Zn concentrations as depicted in Figure 58. In general, substrate formulations containing coffee spent-ground were adequate for *G. frondosa* mycelial growth but did not develop carpophores. It is worth highlighting that coffee spent-ground contains caffeine, tannins, and polyphenols (Fan, Pandey, and Soccol 2001). The toxic nature of these compounds may affect the basidiome development in *G. frondosa* aside from the high zinc content provided by this coffee agribusiness waste.

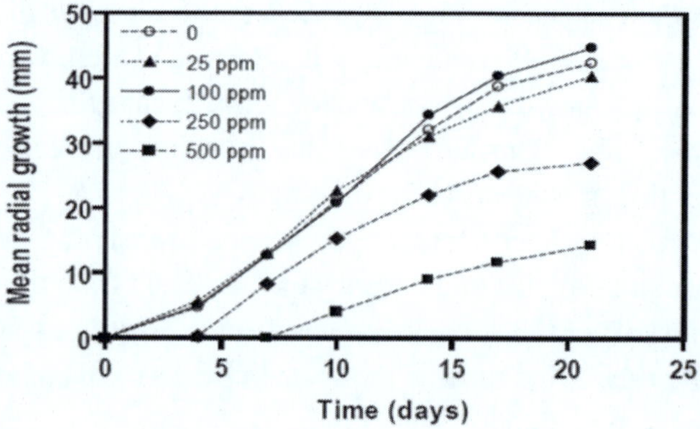

Figure 58. Mean radial growth of *G. frondosa* on PDA medium supplemented with different concentrations of $ZnCl_2$. Source: Montoya, Orrego, and Levin (2012)

MODELING THE SOLID-STATE CULTIVATION OF *G. FRONDOSA*

The cultivation of Maitake (and other mushrooms) in plastic bags to produce fruiting bodies is in itself a solid-state fermentation process (in the biotechnological sense), i.e., a heterogeneous biological transformation of the substrate into cell biomass and different products. During these fermentation processes, the management and control of the operating variables is not simple. The main factors that must be taken into account for solid-state fermentation (SSF) due to its effect on the development of the organism (in this case, the fungal cells) are the substrate pretreatment, particle size, water content of the substrate, type and size of the inoculum, and control of the environmental conditions during the process (Mitchell et al. 2004, Pandey 2003). The relevant environmental conditions for SSF processes are the temperature (removal of metabolic heat), concentrations of carbon dioxide and oxygen in each growth phase, relative humidity, and total time of the process, among others.

One of the main challenges of SSF is maintaining product reproducibility among different processing batches. The heterogeneity of the mixture of solid materials used as substrates, the non-uniform distribution of the different substances and biological materials involved throughout the solid particles, and the complexity of the mass and heat transfer phenomena between and inside the particles, among other factors, provide concentrations of the nutrients required by the mushrooms during their growth and development; the concentrations of metabolites that are synthesized or released into the solid medium vary significantly, thus affecting the growth of the mushrooms, especially when fruiting bodies are being produced. This result explains why these types of SSF processes are very difficult to standardize and control.

Kinetic studies of the fermentation processes may provide valuable insights into the primary regularities and relationships between the process variables. This knowledge is useful for describing (and even predicting) the behavior of the biological system during the growth and development of different organisms on a variety of substrates (Cardona, Sánchez, and Gutiérrez 2010). In this regard, the results of the kinetic studies can be directly applied to design, operate, and control complex SSF processes such as the cultivation of mushrooms. The mathematical modeling of fermentation kinetics plays a crucial role in fulfilling these purposes. Unfortunately, the development of kinetic models for mushroom cultivation is far from the modeling level of conventional submerged fermentations such as those used to produce bioproducts such as ethanol, citric acid, glutamic acid, penicillin, baker's yeast, single-cell protein, etc. Commonly, SSF involves the growth of filamentous fungi on natural solid substrates, in which the carbon source is part of their structure. In most cases, empirical models must be used to monitor what happens during the cultivation processes, resulting in complex submodels due to the heterogeneity of the processes. Empirical models are based on experience and require experimental data from the process to determine their kinetic parameters (generally, the constants involved in the equations). In particular, mushroom fructification is quite difficult to model through mathematical expressions since the entire production cycle of the macromycetes involve different phases (mycelial growth, primordia formation, and development of fruiting bodies) in which each one of them implies different phenomena whose mathematical description is very complex. In fact, many of the transformations that occur during these phases have remained unknown. For this reason, most research efforts on this topic are aimed at modeling the colonization of the substrate by the fungal mycelium during SSF processes. A few studies have been published regarding the modeling of vegetative growth for white rot fungi, such as *Coriolus versicolor*, *Lentinula edodes*, and *Pleurotus ostreatus* (Montoya, Sánchez, and Gutiérrez 2011, Montoya, Sánchez, and Levin 2015). In the case of *G. frondosa*, the published studies are even more scarce. The following research cases address this issue.

Research Case 25. Assessing the Concentrations of G. frondosa in Solid Substrates

One of the main difficulties that arise during solid-state fermentations is the estimation of the biomass, which is essential for kinetic studies. This is due to the strong interaction between the parts of the organism and the substrate, which does not allow them to be separated. This issue is even more complex for filamentous fungi since the hyphae penetrate the particles of the solid material in all directions. It has been suggested that an indirect estimation of the growth of the organism on the substrate could be performed by monitoring a product or substance that differs from all the others found in the solid medium and that can be measured.

The main components of the fungal cell wall are polysaccharides, which correspond to 80-90% of its dry mass. One of these components is chitin, a polysaccharide that makes up part of the fungal skeleton and is responsible for the rigidity and shape of the cell wall. Chitin is the characteristic component of the taxonomic groups Zygo-, Asco-, Basidio-, and Deuteromycetes, and it is absent in other groups, such as Oomycetes. Therefore, the chitin content is characteristic of true fungal species and independent of the variety. Moreover, the chitin levels of fungi can be regulated by the type of nutrition provided to each species (Vetter 2007).

In a previous study (Montoya, Orrego, and Levin 2011), chitin was selected as the compound that differed from all the others found in the given solid medium, and it is a characteristic component of *G. frondosa* fungal biomass. Thus, the chitin content was determined by the glucosamine titration method (Plassard, Mousain, and Salsac 1982). This method quantifies the concentration of N-acetyl-*D*-glucosamine (hereinafter NAG), which is the structural unit of chitin. In this way, the chitin concentration can be determined in the samples of substrates colonized by the filamentous fungi. Thus, the presence of chitin or glucosamine in a substrate-fungus mixture is evidence of the presence of this organism in this system. To determine the biomass content during the different phases of *G. frondosa* growth and development, the glucosamine concentration measurement was used during the different cultivation times on the oak sawdust-based substrate, including the measurement of this substance in the carpophore.

The samples of substrates that were fermented or colonized by *G. frondosa* in bags were subjected to acid hydrolysis with 6 N HCl to release the NAG, which was followed by alkalization and reaction of the alkalinized hydrolysate with reagents prepared from acetylacetone and *p*-dimethylaminobenzaldehyde; the product is measured by spectrophotometry at 530 nm. A standard curve was made with a solution of analytical grade NAG. The biomass content was indirectly determined by measuring the NAG in the solid substrate and recalculated to uncover the dry biomass concentration (in mg of fungal biomass per g of dry substrate) by accounting for the NAG content of *G. frondosa* mycelium. For this purpose, the Maitake was grown by submerged fermentation in 100-mL Erlenmeyer flasks in 25 mL medium containing glucose (30 g/L), yeast extract (6 g/L), $MgSO_4 \cdot 5H_2O$ (0.5 g/L), and K_2HPO_4 (0.5 g/L). To determine the NAG content of the mycelium obtained by submerged fermentation with the same glucosamine titration method as well as the concentration of mycelial biomass by the dry weight method, the NAG mass per unit mass of mycelium is calculated. This value is used to quantify the amount of biomass per gram of dry substrate. Thus, by using this method, several researchers have published the chitin contents of various basidiomycete species (Ivshin et al. 2007, Nitschke et al. 2011). This methodology was applied to obtain the time profile of *G. frondosa* fungal biomass during the solid-state fermentation of oak-based substrates as discussed in the following research case.

Research Case 26. Modeling G. frondosa Fungal Growth during Solid-State Fermentation

In a previous study (Montoya, Orrego, and Levin 2011), a solid-state fermentation process using the mushroom *G. frondosa* was performed. For this process, the two substrate formulations detailed in Research Case 24 were employed. The total fermentation time for the substrate formulation containing coffee spent-ground was of 75 days (no fructification occurred), while for the substrate based on oak sawdust without the coffee waste, the total fermentation time was 107 days since this time included the incubation phase, development of mycelial coat, and formation of fruiting bodies.

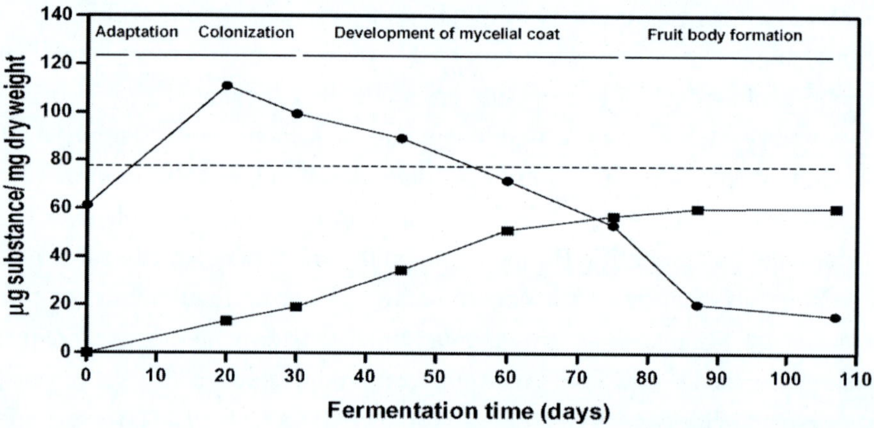

Figure 59. Time course of reducing sugar consumption and NAG formation during the solid-state fermentation of *G. frondosa* on oak sawdust-based substrate. NAG (▲); reducing sugars (●). Growth phases of *G. frondosa* during SSF: adaptation phase (days 1-7), complete colonization of the substrate (days 7-30), development of the mycelial coat (days 30-75), and fruiting body formation (days 76-107; first harvest on 87th day, second harvest on 107th day). Source: Montoya, Orrego, and Levin (2011).

The experimental data required for the mathematical modeling of the Maitake growth and development kinetics for the process using the substrate formulation without coffee spent-ground consisted of the concentration of reducing sugars and the fungal biomass (expressed as NAG, see previous research case). To determine the reducing sugar concentrations, the water-soluble compounds were extracted from the samples by hot water hydrolysis for 3 h, and then the sugars were determined in the filtrate by Somogyi-

Nelson method (Nelson 1944, Somogyi 1945). The substrate samples were collected at different incubation periods (20, 30, 45, 60, 75, 87, and 107 days). The samples were dried at 90°C until reaching a constant weight, ground in a mortar, and stored for chitin determination (including its concentration in thc biomass of the fruiting bodies).

Figure 59 shows the experimental data obtained for the substrate and biomass throughout the total fermentation time (incubation and fructification phases) for Maitake production from a medium based on oak sawdust. In the first fermentation segment (approximately 20 incubation days), a maximum in the amount of reducing sugars is observed, higher than their initial value in the crude substrate. This finding is attributed to the formation of simple sugars as a product of the enzymatic activity of the mushrooms on the starch, cellulose, and hemicellulose contained in the medium. These polysaccharides are used by the fungus to obtain its carbon and energy source needed for its development. Here, it is also corroborated that the crude substrate does not contain NAG. Furthermore, on the 20th day of incubation, there is already NAG formation, which indicates that the mushroom is in its vegetative phase. It is possible to observe that the NAG content in the colonized medium increases to a stable value of 60.5 μg /mg ds (dry sample), a value that is close to but lower than the NAG content in the fruiting body of the obtained Maitake: 77.5 μg /mg ds (indicated with the dotted horizontal line in Figure 59).

The mathematical model of the batch SSF for *G. frondosa* represents a system of ordinary differential equations, each of which describes the behavior of one kinetic variable. In this research, the two studied variables were the substrate (in terms of reducing sugars) and the biomass (in terms of NAG). Several kinetic expressions have been reported to describe solid-state fermentation processes, including the linear model, logistic model, and two-phase model (Sangsurasak, Nopharatana, and Mitchell 1996). Some empirical equations associated with these models are shown in Table 35.

These expressions do not include the effect of the nutrient concentrations on the biomass growth rate. The linear, exponential, and logistic models have been widely employed, while the two-phase model is relatively more recent and involves expressions for the acceleration and deceleration of biomass (Ikasari and Mitchell 2000). In this later model, the exponential phase (equation 5a) is followed by the deceleration phase, the expression of which is represented by equation 5b. This expression is affected by a decrease, which is due to two factors. First, the time at which the deceleration phase begins is time t_a, with parameter L represented as the relationship between the specific growth rates of the deceleration and acceleration phases, respectively. Second, deceleration is described by an exponential decay, as reflected in the specific growth rate with first-order rate constant k. One difficulty in applying this model is that the exponential period is typically short, but it has been indicated when process adjustments are required with little experimental data (Van de Lagemaat and Pyle 2005).

Table 35. Differentiated and integrated forms of some empirical models on biomass growth when applied to solid-state fermentation systems

Model	Differential form	Eq.	Integrated form	Eq.
Linear	$\frac{dX}{dt} = K$	(2)	$X = Kt + X_o$	(6)
Expo-nential	$\frac{dX}{dt} = \mu X$	(3)	$X = X_o e^{\mu t}$	(7)
Logistic	$\frac{dX}{dt} = \mu X\left(1 - \frac{X}{X_m}\right)$	(4)	$X = \frac{X_m}{1 + \left(\frac{X_m}{X_0} - 1\right) e^{-\mu t}}$	(8)
Two phases	$\frac{dX}{dt} = \mu X,\quad t < t_a$	(5a)	$X = X_0 e^{\mu t},\quad t < t_a$	(9a)
	$\frac{dX}{dt} = \left[\mu L e^{-k(t-t_a)}\right] X,\quad t \geq t_a$	(5b)	$X = X_A exp\left[\frac{\mu L}{k}\left(1 - e^{-k(t-t_a)}\right)\right],\quad t \geq t_a$	(9b)

Remarks: X – microbial biomass, t – time, K – linear development rate constant, μ – specific growth rate, X_o – initial biomass concentration, and X_m – maximum amount of possible biomass; t_a, L, and k are parameters of the two-phase model. Source: Montoya, Orrego, and Levin (2011).

The behavior of the data obtained for the NAG concentration as a function of the fermentation time does not fit the linear and exponential models presented in Table 35 (equations 2 and 3), because the biomass curve calculated by these models is always increasing. In contrast, the logistic and two-phase models (equations 4 and 5) do describe a trend with finite growth, which fits the experimental data up to an upper limit represented by the maximum NAG concentration in the fungus (77.5 μg/g ds).

To complete the kinetic model for the behavior of the two variables studied here, it is necessary to include the mathematical expression describing how the substrate (reducing sugars) is consumed during the fermentation process. In this research, two models were proposed, both of which included two kinetic equations for the biomass and substrate. Model I is based on logistic equation (10) for biomass in terms of NAG and equation (11) for the substrate, which shows a direct proportionality between the substrate consumption rate (d*S*/d*t*) and the NAG formation rate (d*G*/*d*t, equivalent to the biomass growth rate) (Weng and Sun 2006):

$$\frac{dG}{dt} = \vartheta G\left(1 - \frac{G}{G_m}\right) \tag{10}$$

$$\frac{dS}{dt} = -\alpha\frac{dG}{dt} \tag{11}$$

where *G* is the NAG concentration (in μg NAG/g ds) that is representative of the mushroom biomass, *S* is the concentration of reducing sugars (in μg of reducing sugars/g ds) that is representative of the substrate (in this case, the main carbon/energy source, not the whole solid medium), *t* is the fermentation time (in days), G_m is the maximum NAG concentration (in μg NAG/g ds), α is a yield coefficient (in μg of reducing sugars/μg NAG) representing the variation in the substrate relative to the biomass increase, and ϑ is the specific NAG formation rate (in day^{-1}) that is equivalent to the specific biomass growth rate (μ). The kinetic parameters α and ϑ were determined from the experimental data (see below), and G_m was taken from the value obtained experimentally for the carpophores (77.5 μg NAG/g ds). Considering that the model can only calculate the decrease in the substrate concentration, the experimental values obtained on the 20th day were used as the initial values (boundary conditions) to solve the system of differential equations (G_0 = 13.33 μg NAG/ g ds and S_0 = 110.85 μg reducing sugars/g ds).

The second model, called Model II or the two-phase model, is based on the equations (5) for the NAG description. For the substrate, this model is based on a linear relationship between the NAG concentration and consumption rate of reducing sugars (-d*S*/d*t*) for the first phase and on an exponential function for the second phase. Thus, Model II includes the following equations:

$$\frac{dG}{dt} = \begin{cases} \vartheta_1 G & t < t_a \\ \vartheta_1 e^{-k(t-t_a)} & t \geq t_a \end{cases} \quad (12)$$

$$\frac{dS}{dt} = \begin{cases} -\propto \vartheta_1 G & t < t_a \\ -\propto \vartheta_1 e^{-k(t-t_a)} & t \geq t_a \end{cases} \quad (13)$$

where ϑ_1 and ϑ_2 are the specific NAG formation rates (in day^{-1}) during the acceleration and deceleration times, respectively, and k is a first-order rate constant; the remaining variables and parameters have the same meaning as they do in Model I. The start of the deceleration phase (t_a) was defined at the 25th day. The kinetic parameters α, ϑ_1, ϑ_2, and k were determined from the experimental data. The determination of all the kinetic parameters for both models was performed using MATLAB® software (MathWorks, USA) by minimizing the sum of the squares in the differences between the experimental data and the data predicted by the models. The values of the parameters adjusted for both models are presented in Table 36. With these values, the systems of equations (10)-(11) corresponding to Model I and equations (12)-(13) corresponding to Model II were solved by considering the initial values of the two variables studied here. The results obtained for Model I are shown in Figure 60a, and for Model II in Figure 60b. The data fitting was quite similar for both models. This represents an advantage for the logistic model compared to the two-phase model, because the former requires only two parameters, while the latter requires four parameters to solve it.

Table 36. Statistical inference in the parameters of two models describing the experimental data of the solid-state fermentation process of *G. frondosa*

Model	Parameter	Value	Confidence interval	Interpretation
Model I (Logistic)	α	1.6715 μg/μg	$1.3951 < \alpha < 1.9479$	The reliable interval is significant (interval does not include zero)
	ϑ	0.0426 day^{-1}	$0.0351 < \vartheta < 0.0501$	The reliable interval is significant (interval does not include zero)
Model II (Two-phase)	α	1.5596 μg/μg	$1.3084 < \alpha < 2.0298$	The reliable interval is significant (interval does not include zero)
	ϑ_1	0.0264 day^{-1}	$0.0178 < \vartheta_1 < 0.0414$	The reliable interval is significant (interval does not include zero)
	ϑ_2	1.6921 day^{-1}	$-1.6564 < \vartheta_2 < 1.7186$	The reliable interval is not significant (interval includes zero)
	k	0.0327	$-0.0131 < k < 0.0448$	The reliable interval is not significant (interval includes zero)

Source: Montoya, Orrego, and Levin (2011).

Considering that there was no great variation in the adjustment of both models, statistical inference was conducted to determine which of the two models best approximates the experimental data of this research (see Table 36).

For the logistic model, the confidence intervals of each of the parameters are significant because they do not pass through zero. In contrast, the parameters ϑ_2 and k are not significant for the two-phase model. The presence of statistically non-significant parameters in a model represents drawbacks in the description of the experimental data and, therefore, it is advisable to remove these parameters from the model. Therefore, it was concluded that to follow and describe the solid-state fermentation process for *G. frondosa* production from an oak sawdust-based culture medium, the two-phase model is reduced to the logistic model.

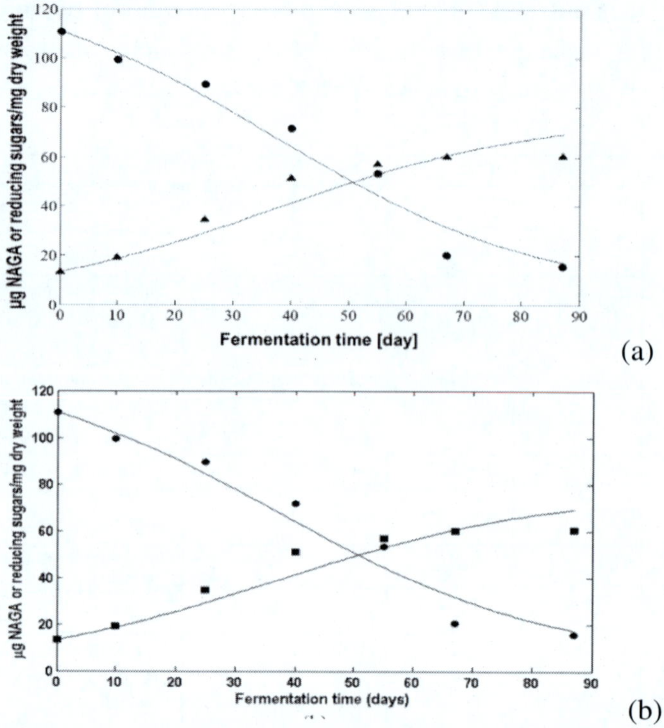

Figure 60. Kinetic modeling results for NAG formation and the consumption of reducing sugars during the solid-state fermentation of *G. frondosa* on oak sawdust-based substrate. The curves are calculated by the models. Experimental data: NAG (▲); reducing sugars (●). (a) Logistic model (Model I). (b) Two-phase model (Model II). Source: Montoya, Orrego, and Levin (2011).

Biologically Active Compounds from *G. frondosa*

G. frondosa (Dickson: Fries) S.F. Gray (Maitake) is a mushroom known for its excellent culinary features in Asia, including its texture, flavor, and aroma. This fungus is highly promising for its medicinal attributes as well. In this regard, mushrooms with medicinal properties, such as *G. frondosa* and *G. lucidum*, have been known for their immunomodulatory properties that occur via the modulation of several cytokines and interleukins (ILs), as confirmed in preclinical and clinical trials (Jin et al. 2016). Moreover, these mushrooms exhibit antitumor and anti-inflammatory properties (see Table

24). Most of these properties are related to the mushroom polysaccharides. The fruiting bodies of the Maitake and the mycelium of the liquid medium contain antitumor polysaccharides (Masuda et al. 2009). Likewise, Maitake polysaccharides have been reported to have antibacterial effects, as well as hypoglycemic (Hong, Xun, and Wutong 2007), hypolipidemic (Harada, Morizono, and Saito 2017), antioxidant (Zhang et al. 2017), hepatoprotective, and hypotensive (Hong et al. 2013) effects, among others.

G. frondosa contains heterogeneous groups of glucose polymers (glucans). In particular, the Maitake contains β-glucans, which consist of a backbone of β(1,3) bonds and β-*D*-glucopyranosyl units with side chains linked by β(1,6) bonds that vary in distribution and length as observed in Figure 61. The β(1,3)-glucans are the major structural components of the fungal cell wall. In the scientific literature, it has been well established that these components refer to their chemical structure but not to their molecular mechanism of action in humans. For example, the immunomodulatory effect of β-glucans depends on the differences in the degree of branching, the length of the polysaccharidic chain, and the tertiary structure of the β-glucans (Borchers et al. 1999). Among the immunostimulatory biopolymers of *G. frondosa*, grifolan, a gel-forming β(1,3)-*D*-glucan with β(1,6)-glycosidic linkages in every third glucopyranosyl residue, should be highlighted (Giavasis and Biliaderis 2006, Ooi and Liu 2000).

Inflammation is a natural process that, when it is persistent and maintained in tissues, can generate biochemical reactions such as the activation of oncogenic cells and metabolic changes, for example, insulin resistance and overactivation of receptor tyrosine kinases (RTKs). In common tumors such as breast and prostate cancers, there are micro areas in the tissues composed of cancer cells that are associated with the adipocytes, macrophages, and fibroblasts that work together to create a microenvironment rich in factors for tumor development, microRNA, and cytokines, all of which have pro-inflammatory properties that promote their survival and chemoresistance (Jin et al. 2016, Rossi et al. 2018).

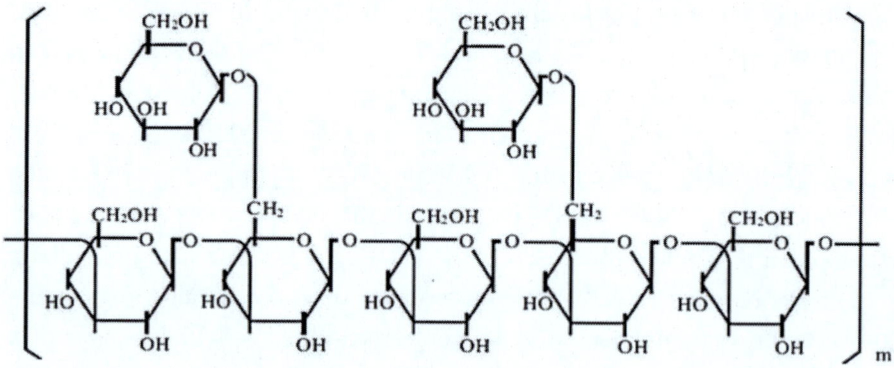

Figure 61. Generic structure of β-glucan polymers showing the backbone of β(1,3)-linked β-*D*-glucopyranosyl units with two β(1,6)-linked side chains every five β(1,3)-linked backbone residues. Source: Rossi et al. (2018).

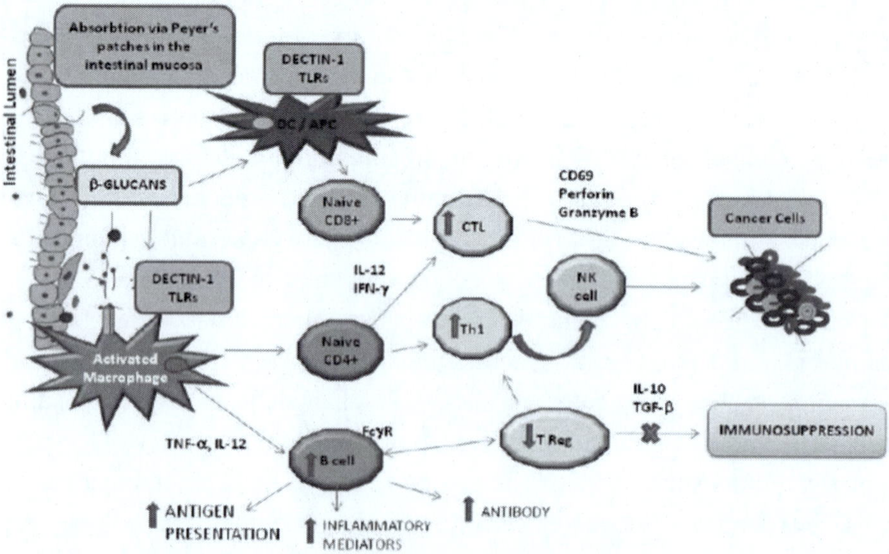

Figure 62. Schematic immunomodulating effect of β-glucans after absorption via intestinal mucosa; β -glucans can directly activate macrophages and dendritic cells (DCs) in the Peyer's patches and can induce both helper T cells (Th) and tumor-specific cytotoxic T cells. In addition, the Fc gamma Receptor (FcγR) provides a critical link between specific humoral responses and cytotoxic T cell lymphocytes (CTCL). Source: Rossi et al. (2018).

The functioning of the immune system is critically involved in the progression of tumors. In this sense, immunotherapy is the main strategy for cancer treatment. Cancer therapy highlights the role of B and T lymphocytes, clusters of differentiation (CDs), natural killers (NK) cells, and mononuclear phagocytic cells. β-glucans from macromycetes exhibit interesting immunoregulatory properties that can be used in cancer treatment. For successful immunotherapy, it is necessary to increase the specific immunity against the tumor and reverse the associated immunosuppression, as illustrated in Figure 62.

The polysaccharides from *G. frondosa* are considered to be effective regulators of the immune system. These polysaccharides activate cells such as macrophages, NK cells, and T lymphocytes, and they promote the secretion of cytokines, including IL-2, IL-8, IL-12, tumor necrosis factor α (TNF-α), and others (Adachi, Nanba, and Kuroda 1987, Kodama, Harada, and Nanba 2002, Masuda et al. 2010).

The oral ingestion of β-glucans from medicinal mushrooms has been found to activate several components of the immune system, including macrophages, NK cells, DCs, and helper T lymphocytes, which affect the viability of tumor cells and enhance the release of several mediators, including lymphokines and interleukins (ILs) (Schwartz and Hadar 2014). The digestive enzymes of mammals do not digest β-glucans; therefore, it is not clear how soluble β-glucans can be absorbed in the gastrointestinal tract. Masuda et al. (2013) noted that the first effect of an orally administered MD-fraction, a highly purified soluble β(1,3)-branched β(1,6)-glucan from *G. frondosa* (see section Maitake market in Chapter 4), was its association with the gut-associated lymphoid tissue (GALT). Likewise, when administered, this fraction takes up antigen-presenting cells (APCs), such as the DCs and macrophages present in the intestinal mucosa, and it adheres in patches. The β-glucans can then be transported by the systemic circulation of the intestinal lumen to the lymphoid tissue. The captured MD-fraction is transported to the spleen. The response of the immune system is thus induced, as demonstrated in mice with tumors.

Kodama, Komuta, and Nanba (2002) conducted investigations in mice with breast cancer by providing a dose of the MD-fraction from Maitake mushrooms for 10 days to evaluate the regression of the tumors. They observed 75-85%-tumor regression after 20 days. Treating with the MD-fraction increased the activity of the NK cells, cytotoxic T cells, macrophages, and delayed-type hypersensitivity T cells by 1.23-2.5 times. Additionally, the production of interleukins 1 (IL-1) from macrophages and interleukins 2 (IL-2) from helper T cells was potentiated. These authors concluded that the ability of immune-competent cells to inhibit tumor growth was enhanced by the MD-fraction. They also conducted human studies as described below:

- A 47-year-old woman with stage II (T_I, N_I, M_0) hepatocellular carcinoma was treated with 80 mg/kg cisplatin (CDDP) four times a day; this treatment was exchanged for 50 mg of MD-fraction and 4 g of whole Maitake tablets per day. After 2.5 years, the tumor had completely disappeared.
- A 54-year-old male with stage III (T_3, N_1, M_0) hepatocellular carcinoma was treated for eight months with 20 mg of adriamycin (ADM), but he refused to continue treatment due to adverse effects. Subsequently, he started treatment with 40 mg of MD-fraction and 5 g Maitake tablets daily. After one year, his bilirubin level decreased from 3.2 mg/dL to 1.6 mg/dL, and his albumin was reduced from 3.1 mg/dL to 1.9 mg/dL.
- A 45-year-old woman with stage III (T_3, N_1, M_o) hepatocellular carcinoma was treated conventionally for one year and experienced severe side effects; her treatment was changed to 4 g Maitake tablets and 100 mg of MD-fraction daily along with the chemotherapy drug 5-fluorouracil (5-FU). In two years, her carcinoma decreased to stage I.

The α-glucan YM-2A that was isolated from *G. frondosa* has been characterized as a highly α(1,6)-branched α(1,4)-glucan, and it also has immunomodulatory properties. Interestingly, YM-2A is more resistant to digestive enzymes than amylopectin and rabbit liver glycogen. Thus, the oral administration of YM-2A can activate macrophages and CDs in GALT. The immunomodulatory effect of YM-2A makes it a promising candidate as an oral therapeutic agent in the translational and clinical research on antitumor immunotherapy (Masuda et al. 2017).

There are many assumptions regarding the aging of humans and animals that have been generated over the years. The free radical theory of aging (oxidative theory) states that free radicals and oxygen are the important factors involved in the phenomenon of biological aging (Nohl 1993). The immune theory of aging states that decreases in the immune functions, especially immune cell function, is the main reason for aging; the most elemental change is the decrease in IL-2 production and the expression of the IL-2 receptor (Liu et al. 1997, Malek and Castro 2010). Several researchers, such as Hong et al. (2013), have performed multiple tests on different fractions of α- and β-glucans from Maitake mushrooms to determine their action as anti-aging agents in animals and humans, finding that these glucans have an anti-aging effect, which is related to their antioxidative and immunomodulatory effects.

Maitake extracts have shown the ability to remove heavy metals. Zhang et al. (2018) conducted tests on a crude product with polysaccharides from dry Maitake mushrooms obtained by hot water extraction and subsequent precipitation with ethanol to evaluate mercury removal capacity in mice that were exposed to this heavy metal. From the extraction processes, the researchers obtained crude fractions of polysaccharides and proteins as well as extracts, mostly of polysaccharides. The contaminated mice were treated with the crude Maitake fractions. Blood samples were then taken from the animals, as well as samples from the livers and kidneys from the same animals after they were sacrificed. The determination of mercury in these samples was performed by using atomic fluorescence spectroscopy. Results indicating 55-88%-mercury removal were obtained from these samples. Therefore, it the ability of Maitake extracts to remove mercury from blood and vital organs such as the kidneys and liver in laboratory mice was demonstrated. This approach is a promising alternative for treating humans who have been exposed to this heavy metal.

The biological properties of Maitake extracts depends on the extraction scheme and methods. In the Bioprocess and Agroindustry Plant at the Universidad de Caldas, ongoing research efforts are being conducted to obtain extracts that exhibit important biological activities. The following research case addresses this topic.

Research Case 27. Extraction Schemes of Polysaccharides from G. frondosa

In the search to obtain polysaccharides from the fruiting bodies of *G. frondosa* mushrooms, several extraction alternatives have been developed to allow for the evaluation of the resulting fractions from the chemical and biological points of view. The overall extraction scheme for obtaining crude fractions of polysaccharides from *G. frondosa* carpophores is depicted in Figure 63, as performed in the Bioprocess and Agro-industry Plant at the Universidad de Caldas. This scheme starts from a dry ground sample of the Maitake fruiting body, which is subjected to a first fat extraction using hexane. Subsequently, the degreased solid is subjected to two aqueous extractions, one followed by the other, at dry fungus:water ratios from 1:20 to 1:40 for 3 h at boiling temperature under reflux. After each extraction, the solid is filtered and an aqueous extract is obtained, from which a sample is taken.

This sample undergoes a subsequent extraction of polysaccharides using 90% ethanol in a 1:4 ratio with subsequent drying and storage. The resulting product is subjected to an evaluation of biological activities. One part of the aqueous extract is collected for lyophilization for the later determination of biological activity. The main extraction procedures for obtaining crude fractions of polysaccharides from the dried, ground fruiting bodies of *G. frondosa* are shown in Figure 64.

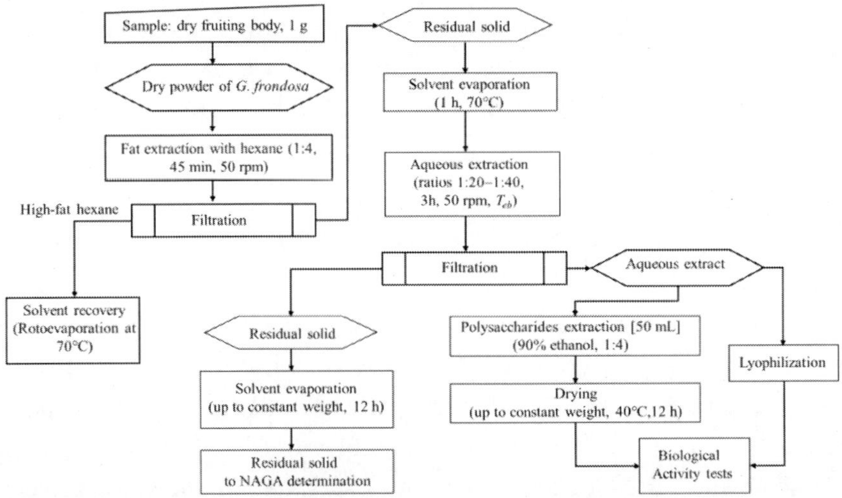

Figure 63. Overall scheme for extracting polysaccharides from *G. frondosa* fruiting body powder to determine the biological activity.

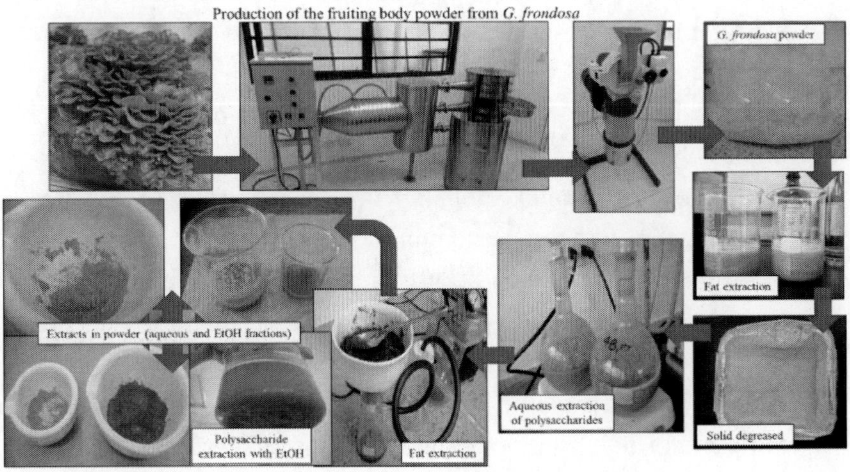

Figure 64. Overall extraction procedures for the production of crude polysaccharide extracts from dry powdered *Grifola frondosa* fruiting bodies.

Currently, there are no conclusive results related to the biological activity of the Maitake fruiting body fractions obtained at the Bioprocess and Agro-industry Plant. In Colombia, there is not a wide availability of laboratories with standardized protocols for biological activity tests in cells or mice. Therefore, in this section, no reference is made to the biological activity tests conducted for this purpose.

FINAL CONSIDERATIONS

The Universidad de Caldas in Manizales (Colombia) has been engaged in research and development for 14 years, studying topics related to the cultivation of macromycetes and their applications. Regarding *G. frondosa* mushrooms, the institution has been involved in related research and development since 2006. Following the execution of different projects, many issues and research challenges have arisen. Some of them are listed, as follows:

- Development of new substrate formulations utilizing lignocellulosic waste or easy-to-grow substitute materials for *G. frondosa* cultivation.
- Standardization of Maitake solid-state culture using new materials in the substrate formulations.
- Evaluation of Maitake quality in regard to the presence of contaminants and competitors in the culture as well as assessments of the environmental conditions for growth and development.
- Evaluation of the influence of the process conditions during the incubation stage on the biological efficiency of the culture.
- Evaluation of the quality of Maitake fruiting bodies obtained from different formulations containing alternative substrates.
- Influence of the feedstocks in the substrate mixtures on the fruiting of *G. frondosa*.
- Relationship between the nutritional content of Maitake carpophores and the nutritional content of the selected substrates.
- Establishment of *G. frondosa* cultivation techniques for tropical weather, in addition to those that were already developed.

- Evaluation of the causes of contamination directly affecting Maitake culture and its relationship to the feedstocks and process conditions.
- Research and development on the use and transformation of spent substrates from Maitake cultivation into valuable products.
- Evaluation of *G. frondosa* fruiting bodies as a functional food.
- Development of new food products using Maitake fruiting bodies.
- Research and development on the extraction schemes of the biologically active substances from *G. frondosa* and on the techniques for determining different biological activities.
- Research and development on the use of Maitake as a nutraceutical product by different groups of human populations to evaluate the influence of the consistent intake of this mushroom.
- Evaluation of the recommended intake doses of Maitake nutraceuticals in different population groups that lead to a change in their status.

REFERENCES

Adachi, K., H. Nanba, and H. Kuroda. 1987. "Potentiation of host-mediated antitumor activity in mice by beta-glutan obtained from *Grfiola frondosa* (maitake)." *Chemical and Pharmaceutical Bulletin* 35:262-70.

Borchers, A. T., J. S. Stern, M. N. Hackman, C. L. Keen, and M. N. Gershwing. 1999. "Mushrooms, tumors and inmunity." *Proceedings of the Society for Experimental Biology and Medicine* 221:281-93.

Cardona, C. A., Ó. J. Sánchez, and L. F. Gutiérrez. 2010. *Process Synthesis for Fuel Ethanol Production*. Boca Raton, FL, USA: CRC Press.

Carlile, M. J., S. C. Watkinson, and G. W. Gooday. 2001. *The Fungi*. London: Academic Press.

Fan, L., A. Pandey, and C. Soccol. 2001. "Production of *Flammulina velutipes* on coffee husk and coffee spent-ground." *Brazilian Archives of Biology and Technology* 44:205–12.

Giavasis, I., and C. G. Biliaderis. 2006. "Microbial polysaccharides." In *Functional Food Carbohydrates*, edited by C. Biliaderis and M. Izydorczyk, 167-214. Boca Raton, FL, USA: CRC Press.

Gilbertson, R. L., and L. Ryvarden. 1987. *North American Polypores Vol. 2. Megasporoporia - Wrightoporia*. Oslo: Fungiflora A/S.

Harada, N., T. Morizono, and M. Saito. 2017. "Blood glucose-reducing and fat-reducing effects of a novel medicinal mushroom, *Grifola gargal* (Agaricomycetes)." *International Journal of Medicinal Mushrooms* 19:1071-81.

Hong, L., W. Weiyu, W. Qin, G. Shuzhen, and W. lebin. 2013. "Antioxidant and immunomodulatory effects of a α-glucan from fruit body of maitake (*Grifola frondosa*)." *Food and Agricultural Immunology* 24 (4):409-18

Hong, L., M. Xun, and W. Wutong. 2007. "Anti-diabetic effect on an α-glucan from fruit body of maitake (*Grifola frondosa*) on KK-Ay mice." *The Jorunal of Pharmacy and Pharmacology* 59:575-82.

Ikasari, L., and D. Mitchell. 2000. "Two-phase model of the kinetics of growth of *Rhizopus oligosporus* in membrana culture." *Biotechnology and Bioengineering* 68:619-27.

Ivshin, V., S. Artamonova, T. Ivshina, and F. Sharnina. 2007. "Methods for Isolation of Chitin–Glucan Complexes from Higher Fungi Native Biomass." *Polymer Science, Serie B.* 49 (11-12):305-10.

Jin, X., J. Ruiz Beguerie, D. M. Y. Sze, and G. C. F. Chan. 2016. *Ganoderma lucidum* (Reishi mushroom) for cancer treatment. *Cochrane Database of Systematic Reviews* 4 (Art. No.: CD007731). Accessed June 2017. doi:10.1002/14651858.CD007731.pub3.

Kodama, N., N. Harada, and H. Nanba. 2002. "A polysaccharides, extract from *Grifola frondosa*, induces th-1 dominant responses in carcinoma-bearing BALB/c mice." *Japanese Journal of Pharmacology* 90:357-60.

Kodama, N., K. Komuta, and H. Nanba. 2002. "Can maitake MD-fraction aid cancer patiens?" *Alternative Medicine Review* 7:236-39.

Lin, D., and Z. Lin. 2014. *Training Course on JUNCAO Technology for Latin America, Caribbean and South-Pacific Regions*. Edited by F. A. a. F. University, Fuzhou, China (Course documentation).

Liu, J., S. Wang, H. Liu, L. Yang, and G. Nan. 1997. "The monitoring biomarker for immune function of lymphocytes in the elderly." *Mechanisms of Ageing and Development* 94 (1–3):177-82.

Malek, T. R., and I. Castro. 2010. "Interleukin-2 receptor signaling: At the interface between tolerance and immunity." *Immunity* 33 (2):153-65.

Masuda, Y., H. Inoue, H. Ohta, A. Miyake, M. Konishi, and H. Nanba. 2013. "Oral administration of soluble β-glucans extracted from *Grifola frondosa* induces systemic antitumor immune response and decreases immunosuppresion in tumor-bearing mice." *International Journal of Cancer* 133:108-19.

Masuda, Y., K. Ito, M. Konishi, and H. Nanba. 2010. "A polysaccharides extracted from *Grfiola frondosa* enhances the anti-tumor activity of bone marrow-derived dendritic cell-based immunotherapy against murine colon cancer." *Cancer Immunology, Immunotherapy* 59:1531-41.

Masuda, Y., A. Matsumoto, T. Toida, T. Oikawa, K. Ito, and H. Nanba. 2009. "Characterization and antitumor effect of a novel polisaccharide from *Grifola frondosa*." *Journal of Agricultural and Food Chemistry* 57:10143-49.

Masuda, Y., Y. Nakayama, A. Tanaka, K. Naito, and M. Konishi. 2017. Antitumor activity of orally administered maitake α-glucan by stimulating antitumor immune response in murine tumor. *PloS one* 12 (3): e0173621. Accessed January 2018. doi:10.1371/journal.pone.0173621.

Mitchell, D. A., O. F. von Meien, N. Krieger, and F. D. Dalsenter. 2004. "A review of recent developments in modeling of microbial growth kinetics and intraparticle phenomena in solid-state fermentation." *Biochemical Engineering Journal* 17 (1):15-26.

Montoya, S., C. E. Orrego, and L. Levin. 2011. "Modeling *Grifola frondosa* fungal growth during solid-state fermentation." *Engineering in Life Science* 11 (3):316-21.

Montoya, S., C. E. Orrego, and L. Levin. 2012. "Growth, fruiting and lignocellulolytic enzyme production by the edible mushroom *Grifola frondosa* (maitake).” *World Journal of Microbiology and Biotechnology* 28:1533-41.

Montoya, S., Ó. J. Sánchez, and L. F. Gutiérrez. 2011. “Representación matemática de la producción de enzimas lignocelulolíticas por fermentación en estado sólido empleando *Coriolus versicolor*.” *Vector* 6:63-70. [“Mathematical representation of lignocellulolytic enzyme production by solid-state fermentation using *Coriolus versicolor*.” *Vector* 6:63-70].

Montoya, S., Ó. J. Sánchez, and L. Levin. 2015. “Production of lignocellulolytic enzymes from three white-rot fungi by solid-state fermentation and mathematical modeling.” *African Journal of Biotechnology* 14 (15):1304-17.

Montoya, S., L. M. Varón, and L. Levin. 2008. “Effect of culture parameter on the production of the edible mushroom *Grifola frondosa* (maitake) in tropical weathers.” *World Journal of Microbiology and Biotechnology* 24:1361-66.

Nelson, N. 1944. “A photometric adaptation of the Somogyi method for the determination of glucose.” *Journal of Biological Chemistry* 153 (2):375-80.

Nitschke, J., H.-J. Altenbach, T. Malolepszy, and H. Mölleken. 2011. “A new method for the quantification of chitin and chitosan in edible mushrooms.” *Carbohydrate Research* 346:1307-10.

Nohl, H. 1993. “Involvement of free radicals in aging: a consequence or causes of senescence.” *Britis Medical Bulletin* 46:653-67.

O’Reilly, P. 2016. *Fascinated by Fungi: Exploring the Majesty and Mystery, Facts and Fantasy of the Quirkiest Kingdom on Earth*. Rhydlewis, Ceredigion, UK: First Nature.

Ooi, V., and F. Liu. 2000. “Immunomodulation and anti-cancer activity of polysaccharide-protein complexes.” *Current Medicinal Chemistry* 7:715-29.

Pandey, A. 2003. “Solid-state fermentation.” *Biochemical Engineering Journal* 13:81–84.

Plassard, C., D. Mousain, and L. Salsac. 1982. "Estimation of mycelial growth of *basidiomycetes* by means of chitin determination." *Phytochemistry* 21:345-48.

Rossi, P., R. Difrancia, V. Quagliarello, E. Savino, P. Tralongo, C. L. Randazzo, and M. Berretta. 2018. "B glucans from *Grifola frondosa* and *Ganoderma lucidum* in breast cancer: an example of complementary and integrative medicine." *Oncotarget* 9 (37):24837-56.

Sangsurasak, P., M. Nopharatana, and D. A. Mitchell. 1996. "Mathematical modeling of the growth of filamentous fungi in solid-state fermentation." *Journal of Scientific and Industrial Research* 55:333-42.

Schwartz, B., and Y. Hadar. 2014. "Possible mechanims of action of mushroom-derived glucans on inflammatory bowel disease and associated cancer." *Annals of Translational Medicine* 2:19.

Somogyi, M. 1945. "A new reagent for the determination of sugars." *Journal of Biological Chemistry* 160:61-68.

Van de Lagemaat, J., and D. L. Pyle. 2005. "Modelling the uptake and growth kinetics of *Penicillium glabrum* in a tannic acid-containing solid-state fermentation for tannase production." *Process Biochemistry* 40:1773-82.

Vetter, J. 2007. "Chitin content of cultivated mushrooms *Agaricus bisporus*, *Pleurotus ostreatus* and *Lentinula edodes*." *Food Chemistry* 102:6-9.

Weng, X.-Y., and J.-Y. Sun. 2006. "Kinetics of biodegradation of free gossypol by *Candida tropicalis* in solid-state fermentation." *Biochemical Engineering Journal* 32:226-32.

Zhang, C., G. Zheng, J. Zhang, C. Hu, X. Sun, R. Chengbo, and L. Jia. 2017. "Antioxidant, antibacterial and anti-aging activities of intracellular zinc polysaccharides from *Grifola frondosa* SH-05." *International Journal of Biological Macromolecules* 95:778-87.

Zhang, W., X. Jiang, S. Zhao, X. Zheng, J. Lan, H. Wang, and T. Bun. 2018. "A polysaccharides-peptide with mercury clearance activity from dried fruiting bodies of maitake mushroom *Grifola frondosa*." *Scientiphic reports* 8 (1):17630.

Chapter 8

ENZYMES FROM *GANODERMA LUCIDUM* AND *GRIFOLA FRONDOSA*

INTRODUCTION

Of all the functions of proteins, catalysis is perhaps the most important; catalysis is performed by special proteins: enzymes. These biological catalysts of proteic origin make the vast majority of the metabolic transformations that occur in all living beings possible. The catabolic processes require the breakdown of complex molecules to obtain energy and the building blocks necessary to synthesize widely diverse compounds that, in turn, are used by the cells to form new biomass and a huge variety of metabolites. White rot fungi have the ability to decompose a wide range of the lignocellulosic materials that are available on Earth to grow, develop, and reproduce. For this purpose, hydrolytic and oxidative enzymes are synthesized to degrade the high molecular weight biopolymers contained in these materials. In this way, the synthesis of lignocellulolytic and other enzymes is a crucial biological process of fungi. These enzymes have attracted considerable interest due to their applications in different sectors, such as the food, feed, chemical, agro-industrial, and biotechnological industries, among others.

Ganoderma lucidum and *Grifola frondosa* are white rot fungi that synthesize and release a substantial amount of lignocellulolytic enzymes into the medium where they grow. In particular, these mushrooms produce cellulases, hemicellulases, and ligninases with potential industrial applications. Moreover, these enzymes play a role that is of paramount importance in degrading substrates during the solid-state cultivation processes required to produce their fruiting bodies. In this sense, the study of these enzymes during the growth and fructification of *G. lucidum* and *G. frondosa*, as well as new methods for their production, represent a strong point of interest considering the development of the global mushroom industry. This chapter presents these topics through several research cases that were conducted by the authors of this book.

LIGNOCELLULOLYTIC ENZYMES

Generalities of Enzymes

Enzymes are synthesized inside cells, but their actions may be performed inside or outside of cells. The first type of enzymes, intracellular enzymes, are responsible for all the metabolic reactions that keep the different biological systems supporting the living functions in balance. These intracellular enzymes can be obtained by chemical or physical methods from the different parts of the organisms that contain them, as required. In addition, modern techniques for the separation and purification of the enzymatic fractions obtained from the extraction processes, including such procedures as centrifugation, ultrafiltration, and chromatography, are being developed to recover and purify these enzymes for various industrial purposes (Blanch and Clark 1997, Bommarius and Riebel 2004). The main function of extracellular enzymes is to reduce the size of the complex molecules found in the environment where the organisms grow and develop. The smaller molecules are subsequently absorbed as sources of nutrients and energy by the organisms that excrete these enzymes (Manavalan, Manavalan, and Heese 2015).

Enzymes that degrade polymers (not necessarily polysaccharides) can employ two attack strategies. They can act on the polymer randomly by effectively fragmenting the molecule into oligomers. The enzymes that display this strategy are called endoenzymes. Alternatively, in the case of polysaccharides, the enzymes can act from one end of the polymeric molecule, generally the non-reducing end, releasing monomers or dimers sequentially. In this case, the enzymes are known as exoenzymes. This action is complemented by enzymes able to act on oligomers or dimers, leading to the complete degradation of the polymer. In this way, the set of all these enzymes (enzymatic complex) acts synergistically (Moore 1998). In general, polysaccharide-degrading enzymes are subject to regulation mechanisms for their synthesis, i.e., they are not formed continuously, but their synthesis is induced by the appropriate substrate and is repressed by easily usable sugars, particularly glucose. The most efficient inductor is the polymer-substrate of the enzymes that will be synthesized. However, due to their high molecular weight, these inductors are compounds that cannot penetrate the cells and exert their effect (Béguin and Aubert 1994).

It is worth highlighting that different forms of the same enzyme are synthesized in slightly different culture media. Despite having similar catalytic features, these enzymes (isoenzymes) differ markedly in their physicochemical characteristics. Multiple forms of isoenzymes may coexist at a given time, but the composition of isoenzymes may also vary throughout the development of the organism that synthesizes them (Carlile, Watkinson, and Gooday 2001). The extracellular enzymes of the cellulolytic and ligninolytic complex are present in multiple forms. The synthesis of isoenzymes may be explained by the expression of different gene products, although they may also be due to posttranslational changes such as proteolysis, glycosylation, or quaternary aggregates. They can be synthesized from differentially regulated genes as a consequence of the different physiological states caused by diverse environmental conditions (the composition of the culture medium, pH, temperature, etc.) (Barr et al. 1996, Blanchette 1995, Carlile, Watkinson, and Gooday 2001, Joselau and Ruel 1994).

Cellulolytic Enzymes

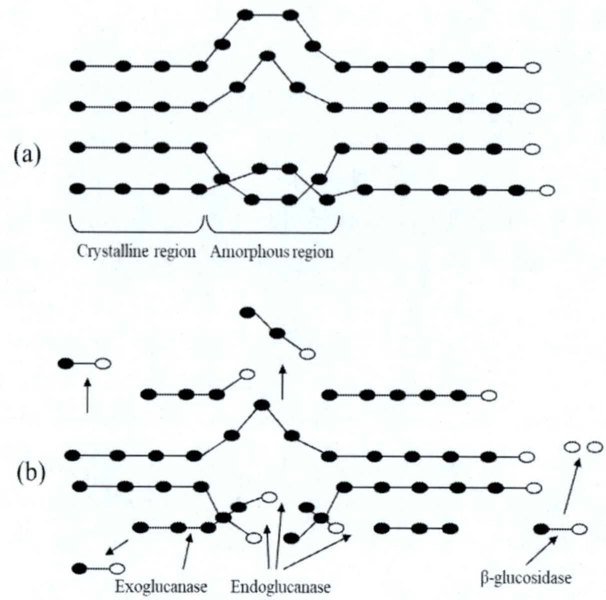

Figure 65. Schematic representation of cellulose action. (a) Crystalline and amorphous regions of cellulose. (b) Action of the cellulase complex on cellulose molecules. Free glucose unit or reducing end of the cellulose chain (○); glucose unit within cellulose or the non-reducing end of the cellulose chain (●).

Cellulose is synergistically attacked by different types of cellulolytic enzymes (cellulases), as depicted in Figure 65. This glucan is initially attacked by endoglucanase, which binds randomly to the cellulose microfibrils in amorphous regions, leading to the hydrolysis of the β(1,4) glycosidic bonds and thus generating multiple attack sites for the exo-β(1,4)-*D*-glucanases. These enzymes act on the non-reducing ends of the cellulose chain, releasing a cellobiose unit. There is continued cooperation between the action of these enzymes and β-glucosidase, which acts on the cellobiose that has been formed or on small oligomers, to produce glucose molecules. The result of the action of these three enzymatic activities is the combined degradation of cellulose. In this way, the complementary activities of these three types of enzymes are considered responsible for synergism, an effect that consists of the action of two or more enzymes in combination being greater than the sum of the individual activities during the degradation process (Papinutti, Dorio, and Forchiassin 2003). Synergism may occur both between and within different classes of cellulolytic enzymes (Barr et al. 1996, Boisset et al. 2000). Although the synergism between the different components of the cellulase complex has not yet been completely elucidated, it is clear that it depends on different factors such as the nature of the substrate, the affinity of the cellulase component for the substrate, the stereo-specificity of said component, the concentration of the enzyme, and the proportion between the enzymatic components (Barr et al. 1996, Mansfield, Mooney, and Saddler 1999).

The synthesis of cellulases is subject to induction by the appropriate substrate (cellulose) and repression by an easily usable carbohydrate source, such as glucose. The most effective inducer is a natural substrate: crystalline cellulose. However, this material is a large polymer that cannot penetrate the cell and induce cellulase synthesis at the transcriptional level. However, there are basal levels of constitutive cellulases that, with a low level of cellulose hydrolysis, would produce some soluble product that is able to penetrate the cell, inducing enzymatic synthesis (Carle-Urioste et al. 1997, Forchiassin and Levin 2007).

Hemicellulolytic Enzymes

The enzymes that degrade the hemicellulose are named according to the substrate on which they act. For example, mannanases degrade mannan, xylanase degrades xylan, etc. (Moore 1998). Among the hemicelluloses, the most abundant is xylan, so xylanases are the most frequently studied enzyme in this group. *D*-xylanases are similar to cellulases, and they are basically composed of two groups of enzymes, endo-β(1,4)-*D*-xylanases (EC 3.2.1.8) and β(1,4)-*D*-xylosidases (EC 3.2.1.37), which take the forms of different isoenzymes in many cases (Goodenough 1996, Liab et al. 2000).

The hydrolysis of these complex molecules requires the interaction of numerous enzymes that cleave the main chain and the lateral ones. The endo-β(1,4)-*D*-xylanase acts randomly on the xylan chain, decreasing its degree of polymerization and releasing xylooligosaccharides, xylobiose, and xylose. The β(1,4)-*D*-xylosidase intervenes after the endoxylanase, acting on the xylooligosaccharides (from the nonreducing ends) or on the xylobiose, forming xylose (Bajpai 1997). However, a complete xylanolytic system requires enzymes such as α-arabinofuranosidase, which acts on the xylose branches and removes the side chains of *L*-arabinose; α-glucuronosidase, which releases glucuronic acids from the side chains, and acetyl xylan esterase, which removes the acetyl substituent groups of xylose (Manavalan, Manavalan, and Heese 2015).

Lignin-Modifying Enzymes

The lignin-modifying enzymes, which are commonly known as ligninases, are oxidoreductases that degrade the complex lignin molecule (see Chapter 1) through an oxidative mechanism. In this sense, these enzymes are not hydrolases like the cellulases and hemicellulases, so the name "lignin-modifying enzymes" has been proposed. The lignin-modifying enzymes can be grouped into peroxidases and phenoloxidases. The first group includes lignin peroxidase (LiP), manganese peroxidase (MnP), and versatile peroxidase (VP), while the second group includes laccase-type enzymes.

LiP, MnP, and VP are peroxidases that employ a heme group as a cofactor and hydrogen peroxide as a primary electron acceptor. LiP can act on a wide range of aromatic phenolic and nonphenolic compounds, including veratryl alcohol (3,4-dimethoxyphenol, a secondary metabolite of some white rot fungi), but it cannot act directly on the lignin molecule, which is too large to fit into the active site (Montoya 2012). Nevertheless, lignin can be broken down in the presence of veratryl alcohol since the free radical that is formed when this enzyme acts on veratryl alcohol can diffuse into the lignified cellular wall, where it oxidizes lignin. MnP catalyzes the oxidation of Mn^{+2} into Mn^{+3} in the presence of hydrogen peroxide and H^{+}; in fact, Mn^{+3} is chelated by the organic acids formed by the fungi (Pérez and Jeffries 1992). This Mn^{+3} can directly oxidize phenolic compounds such as lignin

(Tuor et al. 1992). VP oxidizes Mn^{+2} into Mn^{+3}, as MnP does, and it oxidizes nonphenolic compounds such as LiP; in this way, this enzyme has the catalytic properties of both enzymes, and hence its name (Martínez 2002, Martínez et al. 1996). Finally, laccase is an oxidase that utilizes four copper atoms as a cofactor and oxygen as an electron acceptor, thus producing water as a product (Thurston 1994). Laccases oxidize phenolic and nonphenolic substrates to form cationic radicals, quinones, or phenoxy radicals.

ACTION OF LIGNOCELLULOLYTIC ENZYMES IN WHITE ROT FUNGI

White rot fungi are responsible for wood degradation. They are represented by basidiomycetes and ascomycetes and are able to completely degrade lignin until its mineralization, leaving the wood in the attacked areas white due to the presence of the exposed cellulose. These fungi produce large amounts of oxidoreductases, which are responsible for the mineralization of lignin, in addition to having the ability to produce and transform nonphenolic substances into phenolic compounds for easy degradation to obtain smaller aromatic substances from lignin. In addition, white rot fungi have important cellulolytic and hemicellulolytic complexes to complete the degradation of the lignocellulosic biomass. These fungi employ the products from the degradation of the lignocellulosic materials as sources of carbon and energy (mostly, glucose and other monosaccha-rides).

The main purpose of the enzyme secretion system in filamentous fungi such as white rot fungi is the growth and transport of nutrients towards the hyphal apex, which is the active area for the secretion of extracellular enzymes. However, it is not impossible that there could be some secretory activity throughout the rest of the mycelium structure to maintain the metabolic activity (Papinutti, Dorio, and Forchiassin 2003). Some of the extracellular enzymes that are produced by these fungi are soluble and freely dispersed in the fluid film surrounding the hypha, but others are fixed in the space attached to the hyphal wall, to the extracellular matrix or to the substrate itself (Hudson 1986).

Regarding cellulose degradation, a number of auxiliary oxidative enzymatic activities have been found in different species of white rot fungi in addition to the three main groups of cellulases indicated above. These enzymes are not essential for cellulose hydrolysis but can assist in the decomposition of this polymer by performing useful conversions such as the oxidation or removal of end products that can cause catabolic repression or the inhibition of the activity of the main groups of cellulolytic enzymes (Coughlan and Ljungdahl 1988, Markham 1994). When a white rot fungus develops on cellulose, two cellobiose oxidoreductases are produced: cellobiose quinone oxidoreductase (CBQ) and cellobiose oxidase (CBO). These enzymes convert cellobiose into cellobiono-1,2-lactone (a δ-lactone), which in turn can be spontaneously (or through the action of one lactonase) transformed into cellobionic acid, from which glucose and gluconic acid are obtained (Ander 1994). The cellobiose oxidoreductases are able to use a wide variety of electron acceptors and play a crucial role in the regulation of cellobiose and glucose levels, the accumulation of which can inhibit the action of endoglucanase. Thus, CBQ degrades the bonds between cellulose and lignin (Moore 1998). CBO reduces Fe^{+3} along with hydrogen peroxide, generating hydroxyl radicals. These radicals can degrade both cellulose and lignin. Thus, CBO is very important during wood degradation by white rot fungi as well.

In white rot fungi, the lignin-degradation system works in synergy with other enzymatic complexes. Figure 66 shows the interaction of the different hydrolytic and oxidative enzymatic complexes secreted by the hyphae when they act on cellulose and lignin. The cellulases (assisted by enzymes such CBQ and CBO) degrade the cellulose up to glucose, from which hydrogen peroxide (H_2O_2) is obtained when this monosaccharide is oxidized into gluconic acid. This product is required for the action of the ligninases. In addition, the hypha produces another metabolite called glyoxal. This metabolite induces the synthesis of glyoxal oxidase, which also produces hydrogen peroxide in the presence of oxygen. Thus, these H_2O_2-producing enzymes are required for the action of the peroxidases. The presence of hydrogen peroxide in the medium along with ferrous iron (Fe^{2+}) in the heme group, and veratryl alcohol (another metabolite synthesized by the hypha) are required for lignin peroxidase activation. Other H_2O_2-dependent enzymes such as manganese peroxidase are activated in the presence of Mn^{2+} to degrade lignin. However, laccase (dependent only on Cu^{2+}) acts on phenolic and nonphenolic compounds in all directions of the lignin structure to convert them into free radicals and low molecular weight products; this process is performed in a synergistic manner with the peroxidases. The action of these oxido-reductase enzymes results in partial lignin depolymerization, with the cleavages occurring without stereospecificity. These oxidative reactions produce free radicals, which can react with each other, provoking repolymerization. There is a balance between the polymerization and depolymerization processes during lignin degradation. In addition, nonspecific reactions can be initiated that can then follow any pathway without subsequent enzymatic control, which has been called "enzymatic combustion" (Kirk and Farrel 1987, Manavalan, Manavalan, and Heese 2015).

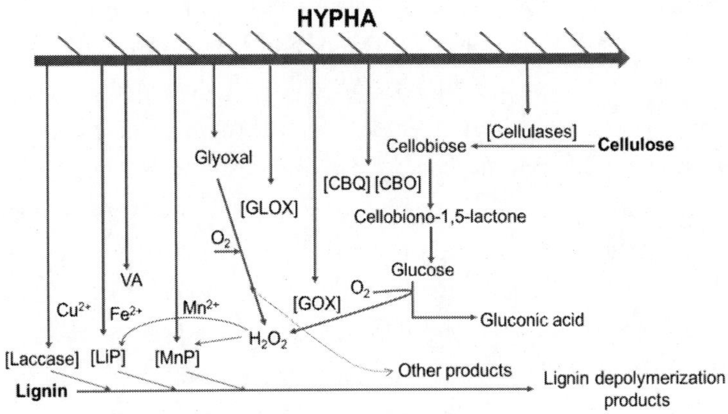

Figure 66. Schematic representation showing the action of the lignocellulolytic system. [CBQ]: Cellobiose quinone oxidoreductase; [CBO]: cellobiose oxidase; [Glox]: glyoxal oxidase; [GOX]: glucose oxidase; [GLOX]: glyoxal oxidase; [MnP]: manganese peroxidase; [LiP]: lignin peroxidase; and VA: veratryl alcohol.

EVALUATION OF LIGNINOLYTIC ENZYMES IN REISHI AND MAITAKE MUSHROOMS

The study of the ligninolytic enzymes from *G. lucidum* and *G. frondosa* is crucial to understanding the mechanism by which these mushrooms degrade the complex biopolymers contained in the substrate used for their growth and development. Knowledge of these mechanisms may allow for improvements in the cultivation process in terms of the enhanced utilization of the given feedstocks as well as the production of higher quality fruiting bodies. For these reasons, in the Bioprocess and Agro-industry Plant at the Universidad de Caldas, several studies and projects have been conducted with the aim of studying and evaluating the lignocellulolytic enzymes from Reishi and Maitake mushrooms. The following research cases exemplify that work.

Research Case 28. Evaluation of Lignocellulolytic Enzyme Production by G. frondosa and Its Relationship with Growth and Fruiting

In a previous study (Montoya, Orrego, and Levin 2012), the following lignocellulolytic enzyme activities of Maitake mushrooms were evaluated when grown on two different substrate formulations: endoglucanase, exoglucanase, β-glucosidase, laccase, manganese peroxidase and lignin peroxidase. It has been well documented that the substrate composition influences the pattern of enzyme production by white rot fungi. Therefore, the ability of a particular substrate to induce or increase the production of lignocellulolytic enzymes is another factor that indirectly improves the ability of those fungi to grow and fructify (Lechner and Papinutti 2006).

In this study, the enzymes associated with the degradation of the components of the solid substrates employed during *G. frondosa* cultivation were quantitatively assessed. The two substrate formulations corresponded to those presented in Research Case 24, i.e., one substrate formulation based on oak sawdust and another formulation based not only on oak sawdust but also on an alternative feedstock, namely coffee spent-ground; corn bran was present in both formulations as the main nitrogen source. Substrate samples (10 g) were collected from both formulations after different incubation periods during *G. frondosa* vegetative growth (on days 20, 30, 45, 60 and 75) and on the 87th and 107th days (fructification cycles) in oak sawdust/corn bran (the unique substrate in which the mushroom fully developed its carpophores). The fungal biomass was estimated by using the content of N-acetyl-*D*-glucosamine (NAG) as previously noted in Research Case 25; the reducing sugars were determined as indicated in Research Case 26, where the procedure used for processing the samples is also described.

The different lignocellulolytic activities were measured for each sample as follows (Montoya, Orrego, and Levin 2012). Cellulolytic and xylanolytic activities were assayed at 50°C, while the ligninases were determined at 30°C. The laccase activity was determined by the oxidation of 0.5 mM ABTS [2,2'azinobis(3-ethylbenzothiazoline-6-sulfonate] in 0.1 M sodium acetate buffer with a pH of 3.6. The manganese-peroxidase (MnP) activity was determined by measuring the oxidation of phenol red (0.01%). The reaction product was measured at 610 nm. The reaction mixture contained 0.05 M sodium succinate buffer pH 4.5, 0.1 mM $MnSO_4$, and 0.1 mM H_2O_2. The lignin peroxidase (LiP) activity was assayed using Azure B as a substrate. The endo-β(1,4)-*D*-glucanase, exo-β(1,4)-*D*-glucanase, endo-β(1,4)-*D*-xylanase, and amylase activities were determined by measuring the reducing sugars produced following the hydrolysis of the substrate by using the Somogyi-Nelson method (Nelson 1944, Somogyi 1945). Amylases are enzymes that hydrolyze starch molecule to form diverse products, including dextrin and progressively smaller polymers made up of glucose units. Measurements were made in 0.1 M sodium acetate buffer at pH 4.8 by using the following substrates: 0.5% carboxymethylcellulose for endoglucanase; 1% crystalline cellulose for exoglucanase; 0.2% xylan from oat spelt, 0.2% for endoxylanase, and 2% soluble starch for total amylase activity. The β-glucosidase activity was determined by measuring the product released from the hydrolysis of 0.02% *p*-nitrophenyl-β-*D*-glucopyranoside in sodium acetate buffer with a pH of 4.8. The reaction was stopped by adding Clark and Lubs buffer (pH 9.8), and the absorbance was measured at 430 nm. Standard curves with glucose (for endoglucanase, exoglucanase, and amylase) and xylose (for endoxylanase) were generated to estimate the enzyme activities, which were expressed in International Units (U), as the amount of enzyme needed to release 1 μmol of product per min. The enzymatic activities in the crude extracts recovered from the solid-state cultures were reported in U/mL. In terms of production, the activity was defined as U per g of dry residue (substrate plus mycelium). For the qualitative LiP determination (Levin, Papinutti, and Forchiassin 2004), *G. frondosa* was inoculated on agar plates containing 12.7 g/L malt extract, 10 g/L glucose, and 20 g/L agar (MEA solid medium) supplemented with 50 μM Azure B. The inoculum consisted of a 25-mm^2 surface agar plug from a 7-day-old culture grown on MEA. Uninoculated plates served as controls for abiotic decoloration. The plates were incubated at 28°C for 28 days. A decolorized zone appeared when the fungus degraded the Azure B dye.

The results obtained for the time profile of the enzymatic activities as well as the biomass (in terms of NAG) and reducing sugars are shown in Figure 67; in addition, the consumption of the primary components of both substrate formulations is presented as well. The substrates were completely colonized after 25-30 days. The reducing sugars produced by the hydrolysis of the enzymatic substrate reached their highest value at the 20th incubation day and then decreased sharply for both formulations. For the formulation without coffee spent-ground, the NAG content increased steadily during the first 87 days, reaching a maximum coinciding with the first fruiting of the 60.5 μg/g dry sample (ds), as noted in Research Case 26. *G. frondosa* almost ran out of reducing sugars during the approximately 85 days of fermentation (see Figure 67a). Since the mushroom growth also ceased at this time, it is suggested that the decrease in the available carbon source limits additional growth.

Many edible mushrooms develop basidiomes when different byproducts or residues from coffee agribusiness are used for cultivation (Fan, Pandey, and Soccol 2001, Job 2004, Velázquez-Cedeño, Mata, and Savoie 2002, Wong and Wang 1991).

Although the substrate formulated with coffee spent-ground was suitable for *G. frondosa* mycelial growth (maximum NAG concentration obtained: 46 μg/g dry substrate at the 75th incubation day, see Figure 67b), it did not develop carpophores (for the corresponding discussion, see Research Case 24).

In this research, although the production of laccase and MnP by Maitake mushrooms was similar in both analyzed substrates, the degradation percentages varied with each substrate. The higher hydrolase activities (both xylanolytic and cellulolytic) were consistent with the Maitake's ability to grow more rapidly in the oak substrate than in the medium supplemented with coffee spent-ground; this finding was consistent with a greater loss of dry weight as well as with the consumption of lignin and polysaccharides. In addition, it is likely that in the substrate based on oak sawdust and coffee grounds, the mushroom received nutrition and energy from the abundant free sugars that were present, which led to a limited use of the cellulose/hemicellulose fraction. The decrease in starch content is probably due to the production of amylolytic enzymes.

Table 37. Average values[a] of the initial content (% on a dry basis) and the decrease (%) in the total dry weight, raw fiber, lignin, hemicellulose, and cellulose content after the solid-state cultivation of *G. frondosa* on oak sawdust-based substrate (107 days) or substrate based on oak sawdust plus coffee spent-ground (75 days)

Substrate	Dry weight	Lignin		Water extractives		Hemicellulose/cellulose	
	Initial content	Initial content	Decrease	Initial content	Decrease	Initial content	Decrease
Oak	46.6±2.5	18.0±1.2	66.9±5.4	22.0±2.2	59.1±4.5	34.0±3.1	44.0±2.8
Oak/coffee	28.3±2.7	12.9±1.1	50.2±3.9	28.2±2.4	43.0±4.1	26.1±2.1	37.3±1.5

Remarks: Water-soluble compounds were extracted after hot water hydrolysis. To determine the acid-insoluble lignin content, the carbohydrates (cellulose and hemicellulose) in the substrate were hydrolyzed and solubilized with sulfuric acid.

[a] Values represent the mean of ten replicates ± standard deviation.

Source: Montoya, Orrego, and Levin (2012).

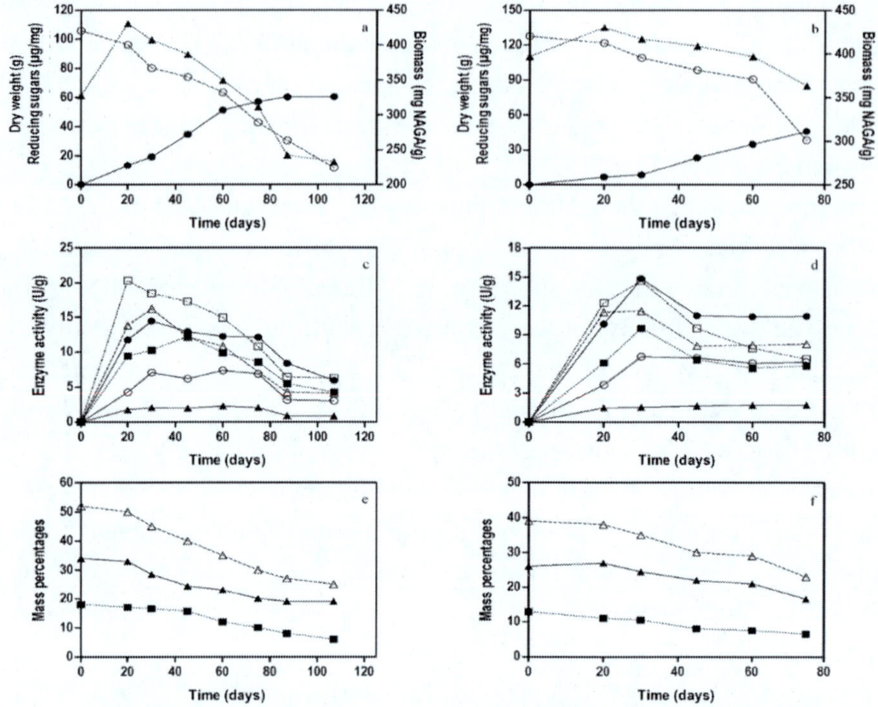

Figure 67. Time profile of fungal growth, lignocellulolytic enzyme production, and fiber, lignin, and carbohydrate contents during the solid-state fermentation of *G. frondosa* on two substrate formulations. (a, b) Biomass (filled circles); dry weight (open circles); reducing sugars (filled triangles). (c, d) Endoglucanase (filled squares), exoglucanase (open triangles), β-glucosidase (filled triangles), endoxylanase (open squares), laccase (filled circles) and manganese-peroxidase (open circles). (e, f) Mass percentage of fiber (open triangles), lignin (filled squares), and carbohydrates (cellulose plus hemicellulose, filled triangles). Substrates: (a, c, e) formulation based on oak sawdust plus corn bran; (b, d, f) formulation based on oak sawdust and coffee spent-ground plus corn bran. Source: (Montoya, Orrego, and Levin 2012).

G. frondosa showed the ability to degrade both substrates, causing a decrease of 67 and 50% in the lignin content and 44 and 37% in the polysaccharides (hemicellulose and cellulose) for the formulations without and with coffee spent-ground, respectively (see Table 37). White rot fungi such as *G. frondosa* are active producers of cellulases and xylanases. The oxidation of lignin does not produce a net energy gain, so lignin is not a substrate in the primary metabolism. The physiological importance of lignin biodegradation is the disarrangement of the lignocellulosic matrix so that fungi can obtain better access to hemicellulose and cellulose substrates. The lignin-degrading systems of white rot fungi are activated mainly during the secondary metabolism of the mushroom. However, *G. frondosa* cultivated by solid-state fermentation produced oxidases along with hydrolases during the trophophase as well, i.e., during the phase of mushroom active growth in which primary metabolites are formed (see Figures 67c and 67d). The cellulolytic, xylanolytic, and ligninase activities were detected in extracts recovered from the solid-state cultures. Although *G. frondosa* tested positive for Azure B plate degradation, a qualitative assay for LiP, attempts to quantify this activity during its solid-state cultivation were unsuccessful. If this enzyme is produced by this mushroom on these substrates, then the LiP levels may be too low to be detected. However, the extraction procedure may not have been adequate to guarantee the recovery of all the enzymes adsorbed on the substrate.

On the oak-based substrate, as shown in Figure 67c, all the enzymatic activities increased during the colonization phase, reaching a peak at approximately days 20-30. Then, they gradually decreased with a very noticeable fall during the formation of fruiting bodies. The addition of coffee spent-ground not only decreased the Maitake growth (Figure 67b) but also the production of endoxylanase, endoglucanase, and exoglucanase (Figure 67d). This reduction can be attributed to the toxic compounds present in coffee spent-ground such as caffeine, tannins, and polyphenols or even to their high zinc concentration. Nevertheless, this addition did not affect the synthesis of the other enzymes analyzed here (ligninases and β-glucosidase). Most of the evaluated enzymes showed a similar pattern throughout the culture time. The highest activities were achieved in this medium around days 20-30, and they decreased steadily until days 45-60, remaining almost constant afterwards (with the exception of the endoxylanase that reached its lowest level at the end of the experiment). β-glucosidase reached its highest titers on days 20-30, showing subsequent constant values. The maximum enzymatic activities are indicated in Table 38 for both substrate formulations.

Table 38. Maximum values[a] of biomass (µg NAG/g dry sample) and lignocellulolytic activities (U/g dry sample) by *G. frondosa* grown in two substrate formulations

Substrate	Biomass	ENG	EXG	β-GL	ENX	AML	LAC	MnP
Oak	60.5±4.6 (87-107)	12.3±0.8 (45)	16.2±0.8 (30)	2.3±0.0 (60)	20.3±2.0 (20)	0.3±0.0 (20)	14.4±0.9 (30)	7.4±0.3 (60)
Oak/coffee	46.0±3.4 (75)	9.7±0.8 (30)	11.4±1.2 (30)	1.7±0.1 (60)	14.6±1.3 (30)	0.2±0.0 (20)	14.8±0.6 (30)	6.7±0.4 (30)

Remarks: The figures in parentheses indicate the day when the maximum activity was reached.
Enzyme activities: ENG – endoglucanase, EXG – exoglucanase, β-GL – β-glucosidase, ENX – endoxylanase, AML – total amylase, LAC – laccase, MnP – manganese peroxidase.
[a] Values represent the mean of ten replicates ± standard deviation.
Source: Montoya, Orrego, and Levin (2012).

The beginning of fructification is verified after a stimulus, such as a physiological stress occurring at that time. Similar to the results obtained in this study, Xing et al. (2006) reported an increase in laccase during the colonization phase, with an activity peak when the substrate was completely colonized and a subsequent laccase decrease during the fructification phase; these authors used substrates prepared with beech sawdust plus wheat and corn flour during the solid-state cultivation of the Maitake. Ligninolytic enzymes have also been repeatedly associated with the production of fruiting bodies in higher basidiomycetes, but the species appear to differ during times of high enzymatic production; there is no clear picture of the functional importance of the enzymes during fruiting (Kües and Liu 2000). In *G. frondosa*, lignocellulolytic enzymes appear to be produced for the growth of the hyphae during the vegetative phase, but not for the development of fruiting bodies. Similarly, elevated ligninase activities were observed during the vegetative growth phase on lignocellulosic substrates, with drastic reductions during the period of fruiting body formation in *Pleurotus ostreatus* (Elisashvili, Kachlishvili, and Penninckx 2008), *Lentinus tigrinus* (Lechner and Papinutti 2006), and *L. edodes* (Mata and Savoie 1998, Ohga and Royse 2001).

The enzymatic activity of *G. lucidum* has also been investigated at the Bioprocess and Agro-industry Plant at the University of Caldas. Besides having importance due to its remarkable medicinal attributes, this white rot fungus is relevant because of its capacity to degrade diverse organic materials via its enzymatic machinery. In the following research case, an enzyme screening that was performed on four species of white rot fungi to evaluate their capacity for degrading lignocellulosic materials is described.

Research Case 29. Determination of Lignocellulolytic Enzymatic Activities for Evaluating the Degradation of Lignocellulosic Materials by G. lucidum and Other Mushrooms

In Research Case 3 (see Chapter 2), ten species of white rot fungi were assessed to estimate their ability to produce lignocellulolytic enzymes as a measure of their lignocellulosic degradation capacity. That assessment was performed through several biochemical tests conducted in Petri dishes using agar-based selective media. From the ten organisms studied in that research case, the four species with the best responses (*Lentinula edodes*, *Coriolus versicolor*, *Ganoderma lucidum*, and *Trametes trogii*) were selected for the subsequent evaluation of their lignin and cellulose degradation capacity, as well as for the quantitative determination of the main enzymatic lignocellulolytic activities. For this purpose, these mushrooms were cultivated on a solid substrate containing 98% oak sawdust and 2% calcium carbonate (percentages on a dry basis) with 60% moisture and a pH of 5.5. The substrate was packed in 125-mL bottles with 30 g of substrate and sterilized at 121°C for 30 min. Then, the substrates were inoculated with each of the strains (four bottles per species) and incubated for 30 days at 25°C under penumbra. After the inoculation process, the bottles were covered with interlon and paraffin film to filter the air during the incubation time. This experiment was performed in triplicate. The determination of the fiber was made from the measurement of neutral detergent fiber, acid detergent fiber, and acid detergent lignin, from which the contents of soluble fiber, hemicellulose, cellulose, and lignin were calculated.

A one-way analysis of variance was performed on the data obtained from the degradation of the fiber components for each one of the four fungal species, yielding statistically significant differences in the degradation of lignin among the mushrooms (p-value < 0.05). *L. edodes* degraded the highest percentage of lignin (51.8%), corresponding to a decrease in the amount of lignin in the raw substrate from 29.9% down to 14.4% after 30 days of incubation, followed by *T. trogii*, which left the substrate with 14.94% lignin during this same time. In addition, a multiple comparison test was performed among the four fungal species to find the final lignin concentration in the substrates, confirming that *L. edodes* was the species that degraded the highest amount of lignin over 30 days with a 95% level of significance.

Regarding the cellulose degradation, there were statistically significant differences (p-value = 0.0362) among the four fungal species, which was confirmed by a multiple comparison test at a 95% significance level. *C. versicolor* degraded the highest percentage of cellulose (22%) corresponding to the decrease in the amount of cellulose in the raw substrate from 59. 6% down to 46.5% in the substrate after 30 days of incubation, followed by *G. lucidum*, which reduced the cellulose content down to 46.95%. However, since all the strains were able to degrade notable percentages of both polymers, any of them could be used for the degradation of lignocellulosic materials. At this point, it is worth highlighting that white rot fungi can enhance the degradation action on fiber components by varying the substrate formulations and environmental conditions. The results obtained in this work are consistent with those of studies by other researchers (Baldrian and Valaskova 2008, Martínez-Anaya et al. 2008).

Table 39 shows the average enzymatic activities (endoglucanase, exoglucanase, β-glucosidase, endoxylanase, laccase, and lignin peroxidase) of the substrates inoculated with the four preselected strains after 30 days of incubation. *C. versicolor* and *T. trogii* exhibited statistically significant differences compared to *G. lucidum* and *L. edodes* in terms of endoglucanase (p-value = 2.7×10^{-5}) and exoglucanase (p-value = 9.2×10^{-5}) activities, so they degraded the highest percentage of cellulose in the substrate. However, regarding β-glucosidase activity, statistically significant differences were found for *L. edodes* and *C. versicolor* (p-value = 0.0144). In the case of the lignin degradation, no significant differences were found in the production of ligninases among the four strains, with *G. lucidum* being a potential organism to use for the degradation of recalcitrant materials with polyphenolic structures or with functional groups similar to those of lignin.

Table 39. Average quantitative enzymatic activities per g of dry substrate (ds) inoculated with four white rot fungi after 30 days of incubation

Enzyme/Strain	*G. lucidum*	*L. edodes*	*C. versicolor*	*T. trogii*
Endoglucanase	3.872	3.371	9.798	10.514
Exoglucanase	5.132	4.469	9.401	8.148
β-glucosidase	1.896	4.436	3.496	4.024
Endoxylanase	29.643	25.924	18.869	26.234
Laccase	7.050	9.489	8.916	9.478
Lignin-peroxidase (LiP)	0.065	0.011	0.097	0.088

The enzymatic activities are expressed in EU (μmol /g ds/min), and the LiP is in U/mL/min.
Source: Montoya, Sánchez, and Levin (2014).

PRODUCTION OF LIGNOCELLULOLYTIC ENZYMES BY SOLID-STATE FERMENTATION

Lignocellulolytic enzymes are valuable products with many industrial applications. These enzymes may be used by the food industry during the production of flours, breads, biscuits, alcoholic beverages, etc. and for the clarification of juices. In addition, they are also employed for textile manufacturing, drug production, and bioremediation processes, among others. In particular, ligninases may be used for the delignification of some agro-industrial waste with animal feed purposes, the detoxification of wastewater streams polluted with phenolic compounds, the decolorization of liquid effluents containing dyes, etc. In turn, cellulases have a wider range of applications in the food, textile, chemical, and microbiological industries. Specifically, cellulolytic enzymes are used to produce fermentable sugars from lignocellulosic waste and by-products. These sugars are converted into value-added products by fermentation, such as ethanol, organic acids, and single-cell protein, among others in the framework of biorefineries. Most commercial lignocellulolytic enzymes (especially cellulases) currently available on the market are produced by submerged fermentation from culture media containing free sugars, such as glucose or sucrose.

However, solid-state fermentation (SSF) processes for producing diverse substances (including enzymes) have been implemented, especially by using micromycetes fungi cultivated in tray bioreactors such as the Plafractor® (Lüth and Eiben 1999, Suyanarayan and Mazumdar 1998). For example, some cellulases used for industrial purposes may be produced by employing *Trichoderma* spp. as process microorganisms. It is worth highlighting that during the production of valuable bioactive compounds from fruiting bodies, the conventional SSF process involves the use of bags containing substrate for the growth and development of the mushroom in cultivation rooms under controlled conditions, as described in previous chapters.

White rot basidiomycetes are excellent producers of cellulases, xylanases, laccases, and manganese peroxidase from lignocellulosic substrates by SSF. These organisms, in most cases, are slow-growing compared to other biotechnological organisms, such as bacteria, yeasts, and many micromycetes. In this respect, a comparison of the synthesis profiles of these lignocellulolytic enzymes performed to study the life cycles of commercial strains of *G. frondosa*, *P. ostreatus* and *L. edodes* cultivated by solid-state fermentation in bags showed that the laccase and MnP activities were high at the end of the colonization phase, and they decreased rapidly during the reproductive phases. In contrast, the enzymatic activities of cellulases and xylanases were expressed in a fairly stable way during colonization and were then increased during the reproductive phase (Asatiani et al. 2008, Montoya, Orrego, and Levin 2012). Therefore, it is important to study the different fermentation methods used to produce lignocellulolytic enzymes by using mushrooms such as Reishi and Maitake.

The use of the solid-state fermentation of filamentous fungi has led to higher yields during the production of enzymes compared to submerged fermentation as well as a reduced inhibition of growth by the components of the substrates and greater stability of the organisms to temperature and pH changes, and the decreased occurrence of enzyme degradation due to the presence of undesirable proteases (Hölker, Höfer, and Lenz 2004). Interesting studies in this field explore the physiological changes in fungi that are presented during the development of fungal cells within SSF processes. For example, there is an apparent accumulation of glycerol, erythritol, and arabinitol that induce glucoamylase genes, which is probably caused by changes in temperature and water activity under these process conditions. The formation of secondary metabolites of numerous filamentous fungi during SSF is associated with the formation of aerial hyphae and spores at the beginning of the secondary metabolism phase. The formation of these secondary metabolites is only reported in this type of process and not during submerged fermentations (Hölker and Lenz 2005). The presence of these secondary metabolites constitutes another reason why the technological development of SSF is necessary, although there are many research challenges to be addressed in terms of the production of lignocellulolytic enzymes and other substances using filamentous fungi by this cultivation method.

Engineering Challenges of Solid-State Fermentation

The technological developments and engineering aspects of submerged fermentation are many years ahead of solid-state fermentation processes. Because the growth of process organisms during submerged fermentation is performed in a liquid culture medium that provides all the necessary nutrients, the principles of stirred-tank reactors from chemical engineering have been applied in a relatively easy way to this fermentation method. In contrast, the presence of a complex and heterogeneous solid phase along with the particularities of the microbial growth (especially when fungi are used) as well as the constraints imposed by mass and transfer phenomena have limited the understanding and development of SSF.

Several methods for producing lignocellulolytic enzymes by SSF have been reported. Different strains of fungi have been tested to quantify their potential for enzyme production using specific solid media. There are bioreactors and SSF processes available on the market for producing lignocellulolytic enzymes.

However, most industrial production of these enzymes has been performed by submerged fermentation due to the problems related to the scale-up of solid-state bioreactors, such as the control of the process variables. In particular, the heat and mass transfer phenomena in such bioreactors are difficult to guarantee. These issues can be mostly attributed to the non-availability of mathematical models that guarantee a proper representation of these processes with high reliability.

The problems that have arisen during the standardization and scale-up of SSF make it difficult to reproduce results. In particular, the handling of the gradients of the temperature, moisture, and nutrient concentrations creates an important bottleneck. These gradients can be increased during the process, thus generating adverse conditions in bioreactors employed for SSF processes, such as static-bed solid-state bioreactors and bioreactors with intermitting stirring of the solid bed like rotary-drum fermenters. The complex interrelation between these environmental conditions contribute to the difficulty of regulating these operating parameters.

One of the most relevant aspects to take into account during the design and operation of SSF is mass transfer. Unlike submerged fermentation, in which the development of fungal biomass generally takes the form of pellets, the development of fungi on solid particles (and sometimes inside them on the surface of meso- and micropores) is accomplished through hyphal growth, as described above. This situation generates important restrictions for the exchange of oxygen and nutrients during SSF. In this sense, factors such as the particle size and the moisture content of the substrates should be properly defined to favor the intraparticle mass transfer (Lonsane et al. 2005, Mitchell et al. 2004). Thus, the mass transfer must be understood as a heterogeneous interaction between all the molecules and towards all the directions of the substrate (between and within solid particles), but this interaction has remained poorly understood when fungal biomass growth is taking place.

When fungal development takes place under aerobic conditions in the bioreactor, there is a considerable increase in heat production, causing a rapid increase in temperature. This effect is desirable in composting processes but is totally adverse during fermentation processes in bioreactors since a large part of the enzymes produced during SSF can be denatured at the end of the process. In the absence of one free aqueous phase, the resulting heat is very difficult to remove. To remove this metabolic heat, induced aerations are generally used. These aerations must be carefully planned to avoid the excessive loss of water from the solid material or the increase of the water content of the substrate, which can provoke the appearance of other opportunistic organisms.

All these issues represent important challenges for the design and control of more efficient SSF processes. Particularly, it is necessary to reach a better understanding of the biological, physical, and chemical processes that occur during the development of the hyphae of white rot fungi in solid media to clarify more suitable process routes for industrializing the SSF processes (dos Santos et al. 2004, Hölker and Lenz 2005, Lenz et al. 2004).

Novel Bioreactor for Producing Enzymes and Other Biologically Active Substances by Solid-State Fermentation

Different solid-state bioreactors with diverse designs are being used at the pilot scale and commercial level. The diversity of these designs is related to the complexity of the type of processes and engineering challenges mentioned above. In general, static-bed solid-state bioreactors and bioreactors with intermittent stirring of the solid bed are intended for fungal cultivation during the vegetative phase, i.e., for the colonization of the substrate by mycelia. In general, the formation of fruiting bodies is not possible in such types of bioreactors. Instead, they are used to produce enzymes and other metabolites, such as exopolysaccharides, mycelial intrapolysaccharides, terpenoids, etc.

In the Bioprocess and Agro-industry Plant at the Universidad de Caldas, a novel solid-state fixed-bed bioreactor for the production of biosubstances (including lignocellulolytic enzymes and polysaccharides) using white rot fungi was designed, built, and started up. The motivation for implementing such a process in this bioreactor was to obtain competitive yields of these compounds and to broaden the possibilities of adding value when agro-industrial waste and mushrooms are employed. This bioreactor was patented in both Colombia and the United States, and it is described in the following research case, in which *G. lucidum* was one of the test organisms.

Research Case 30. Bioreactor to Obtain Bioactive Substances by Solid-State Fermentation Using G. lucidum and Other Macromycetes Fungi

On 14th October 2014, the Superintendence of Industry and Commerce of Colombia granted an Invention Patent (Resolution 61503) entitled *Biorreactor para la obtención de sustancias bioactivas por fermentación en estado sólido empleando hongos macromicetos* (Bioreactor for the production of bioactive substances by solid-state fermentation using macromycete fungi) to the Universidad de Caldas. The inventors were two of the authors of this book (Drs. Sandra Montoya and Óscar J. Sánchez) along with Dr. Luis. F. Gutiérrez, who are researchers at the Universidad de Caldas. In addition, the United States Patent and Trade Office granted patent US9,556,409 entitled *Bioreactor to obtain bioactive substances through solid-state fermentation using macromycetes fungi* to the Universidad de Caldas by the same inventors on January 31st, 2017 (Montoya, Sánchez, and Gutiérrez 2017).

The patented invention corresponded to a fixed-bed bioreactor with natural convection or forced draught (forced aeration) that possessed removable boxes in which the solid substrate is placed for fungal cultivation. This bioreactor is described briefly, as follows:

- The body of the bioreactor (see Figure 68) is composed of a stainless steel chamber (1) with lateral spaces in a concave shape, a base diffuser (5) for distributing air and to support the boxes containing the substrate, a main lid (7 in Figure 68) provided with another lid that allows for the exchange of air with the outside (with a stainless steel mesh and filter) (29 in Figure 70), three peepholes for inspecting the incubation of the substrates (25 in Figure 70), two samplers with sample-receiving chambers isolated from the external environment (26 in Figure 70), a bottom air diffuser (5 in Figure 68), two air ejectors (2 in Figure 68 and 21, 22 in Figure 69), and an air mixer (3 in Figure 68 and 33 in Figure 70). The two samplers are provided with fastening means, and there is a main container for collecting samples that are in contact with the environment, a main window for collecting samples, a handle for the window controlling the entrance of air into the bioreactor, a valve controlling the entrance of air into the bioreactor, a general sampling container, and an access window for the bioreactor.

- There are two bodies for the lateral air diffuser system (13 in Figure 68), each of which faced combined positive (15 in Figure 68) and negative (15 in Figure 68) spin dispersion for distributing air when working with natural convection and directing the air flow when the bioreactor has forced aeration.

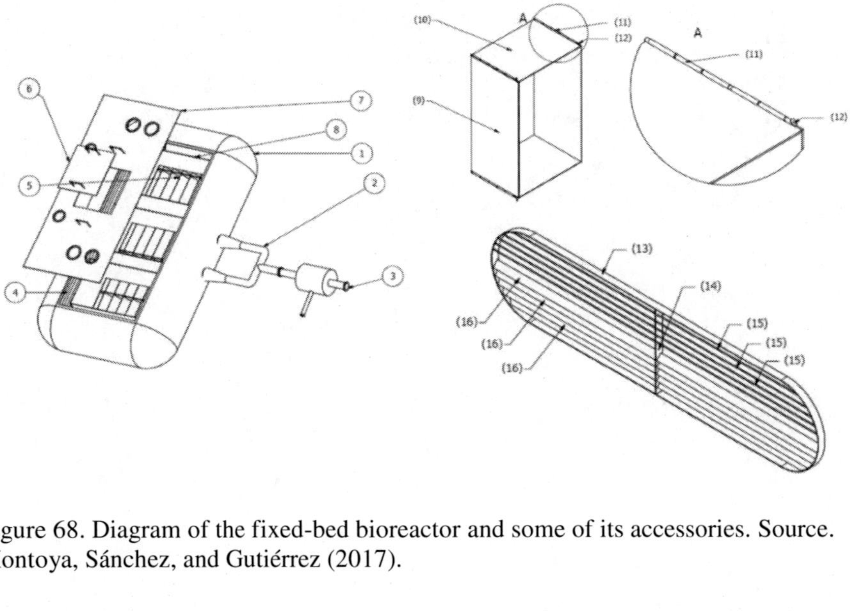

Figure 68. Diagram of the fixed-bed bioreactor and some of its accessories. Source. Montoya, Sánchez, and Gutiérrez (2017).

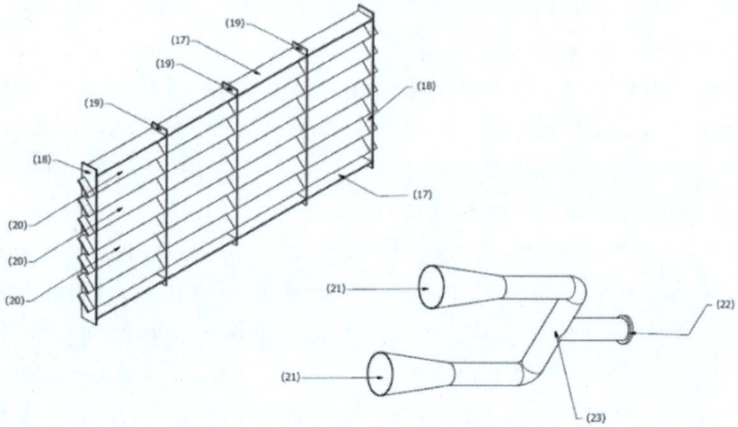

Figure 69. Air diffuser system and air ejectors of the fixed-bed bioreactor. Source. Montoya, Sánchez, and Gutiérrez (2017).

- There are three substrate-containing boxes that can be disassembled (10 in Figure 68). These boxes contain the substrate formulated with lignocellulosic substrates that are inoculated with species of white rot fungi. Each box has a 10-kg capacity and is provided with hinges with a block (11, 12) on each side, which allows for the complete disassembly and release of the substrate block after incubation.

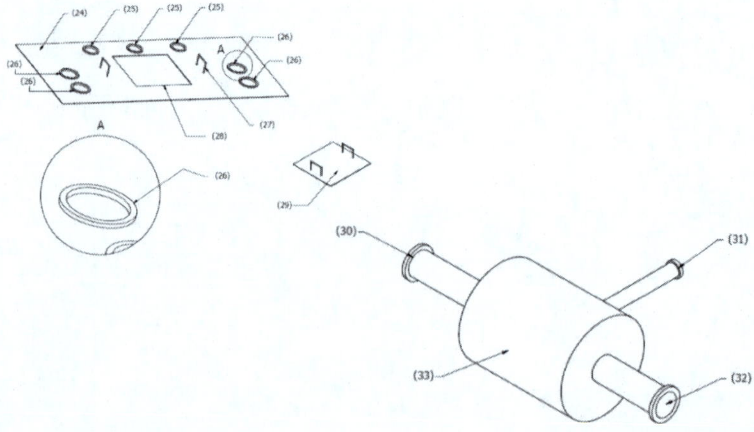

Figure 70. Diagram of the bioreactor lid (top) and air mixer. Source. Montoya, Sánchez, and Gutiérrez (2017).

Figure 71. General view of the fixed-bed solid-state bioreactor with natural convection or forced aeration to produce bioactive substances at the pilot scale, as designed and built at the Universidad de Caldas (Colombia).

- There is a control box for the temperature and relative humidity comprising a thermocouple selector, a display for six temperature points, a relative humidity display, a power-on indicator light, a general switch, and a power-off indicator light.
- There is an air mixer (33) that consists of a fresh air entrance (30 in Figure 70), an air recirculation pipe (31), an air exit towards the bioreactor (32), and a mixing tank for fresh and recycled air (33).
- There are stainless steel pipes with a double air expansion device and a mixer for injecting recirculating air when the bioreactor is working with forced aeration (2 in Figure 68).

The general view of the fixed-bed bioreactor is shown in Figure 71. The main features of this fixed-bed bioreactor are listed below:

- The bioreactor has its own design that allows for operation with natural convection and a forced air draft.
- This equipment was specially designed for the growth and development of white and brown rot macromycetes fungi.

- In the bioreactor, biosubstances may be produced as enzymatic lignocellulolytic complexes; it is possible to enhance the production of some of the enzyme groups contained in these complexes such as cellulases, hemicellulases, amylases, and ligninases.
- In the bioreactor, intra- and exopolysaccharides may be produced; the intrapolysaccharides are contained in the mycelium that colonizes the solid substrate, and the exopolysaccharides are excreted into the medium and carried in the exudate that is formed during solid-state cultivation.
- The colonization that occurs in the bioreactor during the solid-state fermentation under controlled conditions allows for the compaction of the material, which is essential for the subsequent fruiting of the fungi macromycetes; once the solid-state fermentation is completed in the bioreactor and the exudates are collected, the substrate blocks can be moved to cultivation rooms with controlled environmental conditions to allow the fruiting bodies to be fructified.
- Other biosubstances, such as statins, interferons, regulators of blood sugar, and regulators of blood pressure, can be obtained in this bioreactor by selecting the substrate (components of the formulation) and the fungal species as well as the operating conditions for the SSF process.

Several experimental runs have been conducted in this fixed-bed bioreactor involving different white rot fungi, including Reishi mushrooms. The production of different biosubstances was evaluated in the resulting exudates and in the solid material colonized by *G. lucidum* and *C. versicolor* during the SSF of lignocellulosic waste in a previous study (Montoya 2012). The enzymatic activities that were determined in the colonized substrate and the exudates are as follows: endoglucanase activity from 8 to 26 μmol/(g dry sample (ds) × min) for the substrate and between 2 and 9 μmol/(mL × min) for the exudates; exoglucanase activity from 8 to 24 μmol/(g ds × min) for the substrate and 1.5-12 μmol/(mL × min) for the exudates; β-glucosidase activity from 2 to 9 μmol/(g ds × min) for the substrate and 0.7-2.5 μmol/(mL × min) for the exudates; endoxylanase activity from 12 to 25 μmol/(g ds × min) for the substrate and 4-9 μmol/(mL × min) for the exudates; lignin peroxidase activity from 9 to 18 μmol/(g ds × min) for the substrate and 2-7 μmol/(mL × min) for the exudates; laccase activity from 12 to 27 μmol/(g ds × min) for the substrate and 5-12 μmol/(mL × min) for the exudates; and manganese peroxidase activity from 5 to 13 μmol/(g ds× min) for the substrate and 1.5-4 μmol/(mL × min) for the exudates.

It is worth highlighting that the phase of vegetative *G lucidum* growth was reduced by 53% (down to 14 incubation days) compared to the cultivation performed in bags. These results show the enhanced possibility of obtaining lignocellulolytic enzymes with higher titers from reduced cultivation times.

FINAL CONSIDERATIONS

This chapter presents a compilation of different research efforts on lignocellulolytic enzymes from the Bioprocess and Agro-industry Plant at the Universidad de Caldas. These efforts were aimed at evaluating the enzymatic activities, determining the cultivation conditions for obtaining enzymes, and performing the separation and purification of lignocellulolytic enzymes from white rot fungi, including *G. lucidum* and *G. frondosa*. Currently, several research projects are being performed on these topics. In particular, a PhD project is now addressing the production of lignin-modifying enzymes using *Trametes versicolor* in the patented fixed-bed bioreactor. From this study, various research issues have arisen related to supplementation with metals to increase the titers of ligninases, the combination of different aeration regimes (natural convection and diverse air flow rates for forced aeration, including the mixing of fresh and agitated air), and the definition of schemes for the separation and purification of the obtained enzymatic cocktails and their stabilization for industrial applications. These issues will be studied in the future.

Regarding the production of lignocellulolytic enzymes from Reishi and Maitake mushrooms, the following research topics will be addressed at the Bioprocess and Agro-industry Plant:

- Detailed physical chemical and enzymatic activity characterizations of lignocellulolytic enzymes from *G. lucidum* and *G. frondosa*.

- Optimization of the cultivation conditions in the novel fixed-bed solid-state bioreactor for producing enzymes and other bioactive substances by using Reishi and Maitake mushrooms.
- Stability and shelf-life studies of lignocellulolytic enzymes obtained from Reishi and Maitake mushrooms.
- Pilot-scale separation and purification of enzymatic cocktails from *G. lucidum* and *G. frondosa* as obtained by submerged and solid-state fermentation.
- Modeling and simulation of downstream operations for flowsheeting and designing the process for producing lignocellulolytic enzymes from white rot fungi.

References

Ander, P. 1994. "The cellobiose-oxidizing enzymes CBQ and CBO as related to lignin and cellulose degradation — a review." *FEMS Microbiology Reviews* 13 (2-3):297-311.

Asatiani, M. D., E. T. Kachlishvili, T. S. Khardziani, E. M. Metreveli, N. A. Mikiashvili, G. G. Songulashvili, N. D. Tsiklauri, S. P. Wasser, and V. I. Elisashvili. 2008. "*Basidiomycetes* as a source of antioxidants, lectins, polysaccharides, and enzymes." *Journal of Biotechnology* 136S:S717-S42.

Bajpai, P. 1997. "Microbial xylanolytic enzyme system: properties and applications." *Advances in Applied Microbiology* 43:141-94.

Baldrian, P., and V. Valaskova. 2008. "Degradation of cellulose by basidiomycetous fungi." *FEMS Microbiology Review* 32:501-21.

Barr, B. K., Y. L. Hsieh, B. Ganem, and D. B. Wilson. 1996. "Identification of two functionally different classes of exocellulases." *Biochemistry* 35 (2):586-92.

Béguin, P., and J. P. Aubert. 1994. "The biological degradation of cellulose." *FEMS Microbiology Reviews* 13 (1):25-58.

Blanch, H. W., and D. S. Clark. 1997. *Biochemical Engineering*. New York: Marcel Dekker.

Blanchette, R. A. 1995. "Degradation of the lignocellulose complex in wood." *Canadian Journal of Botany* 73:S999-S1010.

Boisset, C., C. Fraschini, M. Schülein, and B. Henriss. 2000. "Imaging the enzymatic digestion of bacterial cellulose ribbons reveals the endo character of the cellobioydrolase Cel6A from *Humicola insolens* and its mode of synergy with cellobiohydrolase Cel7A." *Applied and Environmental Microbiology* 66 (4):1444-52.

Bommarius, A. S., and B. R. Riebel. 2004. *Biocatalysis*. Atlanta: Wiley-VCH.

Carle-Urioste, J. C., J. Escobar-Vera, S. El-Gogary, F. Henrique-Silva, E. Torigoi, O. Crivellaro, A. Herrera-Estrella, and H. El-Dorry. 1997. "Cellulase induction in *Trichoderma reesei* by cellulose requires its own basal expression." *Journal of Biological Chemistry* 272 (15): 10169-74.

Carlile, M. J., S. C. Watkinson, and G. W. Gooday. 2001. *The Fungi*. London: Academic Press.

Coughlan, M. P., and L. G. Ljungdahl. 1988. "Comparative biochemistry of fungal and bacterial cellulolytic enzyme systems." *FEMS Symposium - Federation of European Microbiological Societies* 43:11-30.

dos Santos, M. M., A. S. da Rosa, S. Dal'Boit, D. A. Mitchell, and N. Krieger. 2004. "Thermal denaturation: is solid-state fermentation really a good technology for the production of enzymes?" *Bioresource Technology* 93 (3):261-68.

Elisashvili, V., E. Kachlishvili, and M. Penninckx. 2008. "Lignocellulolytic enzymes profile during growth and fruiting of *Pleurotus ostreatus* on wheat straw and tree leaves." *Acta Microbiologica et Immunologica Hungarica* 55 (2):157-68.

Fan, L., A. Pandey, and C. Soccol. 2001. "Production of *Flammulina velutipes* on coffee husk and coffee spent-ground." *Brazilian Archives of Biology and Technology* 44:205–12.

Forchiassin, F., and L. Levin. 2007. "Enzimas extracelulares. Degradación de polímeros vegetales." In *Introducción al Reino de los Hongos y los Grupos Afines*, edited by E. Albertó and E. Vadell. San Martín, Argentina: Universidad Nacional de San Martín. ["Extracellular enzymes. Degradation of plant polymers." In Introducción to the

Kingdom of Fungi and Related Groups, edited by E. Albertó and E. Vadell. San Martín, Argentina: National University of San Martín].

Goodenough, P. W. 1996. "Structural studies on cellulases, pectinases an xylanases." In *Enzymes for Carbohydrate Engineering*, edited by K.-H. Park, J. F. Robyt and Y.-D. Choi, 83-107. Amsterdam: Elsevier.

Hölker, U., M. Höfer, and J. Lenz. 2004. "Biotechnological advantages of laboratory-scale solid-state fermentation with fungi." *Applied Microbiology and Biotechnology* 64 (2):175-86.

Hölker, U., and J. Lenz. 2005. "Solid-state fermentation — are there any biotechnological advantages?" *Current Option in Microbiology* 8 (3):301-06.

Hudson, H. 1986. *Fungal Biology*. Maryland, USA: Edward Arnolds Eds.

Job, D. 2004. "La utilización de la borra del café como substrato de base para el cultivo de *Pleurotus ostreatus* (Jacq.:Fr.) Kummer." *Revista Iberoamericana de Micologia* 21 (4):195-97. ["Use of coffee spent industry residues for production of Pleurotus ostreatus (Jacq.:Fr.) Kummer in solid state fermentation." *Iberoamerican Journal of Mycology* 21 (4):195-97].

Joselau, J. P., and K. Ruel. 1994. "Wood polysaccharides and their degradation by fungi." In *Host Wall Alterations by Parasitic Fungi*, edited by O. Petrini and G. B. Ouellette, 334-87. St. Paul, MN, USA: APS Press.

Kirk, T. K., and R. L. Farrel. 1987. "Enzymatic combustion: the microbial degradation of lignin." *Annual Reviews of Microbiology* 41:465-505.

Kües, U., and Y. Liu. 2000. "Fruiting body production in basidiomycetes." *Applied Microbiology and Biotechnology* 54 (2):141-52.

Lechner, B. E., and V. L. Papinutti. 2006. "Production of lignocellulosic enzymes during growth and fruiting of the edible fungus *Lentinus tigrinus* on wheat straw." *Process Biochemistry* 41 (3):594-98.

Lenz, J., M. Höfer, J. B. Krasenbrink, and U. Hölker. 2004. "A survey of computational and physical methods applied to solid-state fermentation." *Applied Microbiology and Biotechnology* 65 (1):9-17.

Levin, L., L. Papinutti, and F. Forchiassin. 2004. "Evaluation of Argentinean white rot fungi for their ability to produce lignin-modifying enzymes and decolorize industrial dyes." *Bioresource Technology* 94 (2):169-76.

Liab, K., P. Azadi, R. Collins, J. Tolan, J. S. Kim, and K. E. L. Eriksson. 2000. "Relationships between activities of xylanases and xylan structures." *Enzyme and Microbial Technology* 27 (1-2):89-94.

Lonsane, B. K., N. P. Ghildyal, S. Budiatman, and S. V. Ramakrishna. 2005. "Engineering aspects of solid state fermentation." *Enzyme and Microbial Technology* 7 (6):258-65.

Lüth, P., and U. Eiben. 1999. "*Solid-state fermenter and method for solid-state fermentation.*" US patent 6620614.

Manavalan, T., A. Manavalan, and K. Heese. 2015. "Characterization of lignocellulolytic enzymes from white-rot fungi." *Current Microbiology* 70 (4):485-98.

Mansfield, S. D., C. Mooney, and J. N. Saddler. 1999. "Substrate and enzyme characteristics that limit cellulose hydrolysis." *Biotechnological progress* 15 (5):804-16.

Markham, P. 1994. "Occlusions of septal pores in filamentous fungi." *Mycological Research* 98 (10):1089-106.

Martínez-Anaya, C., E. Balcázar-López, E. Dantán-González, and J. L. Folch-Mallol. 2008. "Celulasas fúngicas: Aspectos biológicos y aplicaciones en la industria energética." *Revista Latinoamericana de Microbiología* 50 (3):119-31. ["Fungal cellulases: Biological aspects and applications in the energy industry." *Latin American Journal of Microbiology* 50 (3):119-31].

Martínez, A. T. 2002. "Molecular biology and structure-function of lignin-degrading heme peroxidases." *Enzyme and Microbial Technology* 30 (4):425-44.

Martínez, M. J., F. J. Ruiz-Dueñas, F. Guillén, and A. T. Martínez. 1996. "Purification and catalytic properties of two manganese peroxidase isoenzymes from *Pleurotus eryngii*." *European Journal of Biochemistry* 237 (2):424-32.

Mata, G., and J.-M. Savoie. 1998. "Extracellular enzyme activities in six *Lentinula edodes* strains during cultivation in wheat straw." *World Journal of Microbiology and Biotechnology* 14 (4):513-19.

Mitchell, D. A., O. F. von Meien, N. Krieger, and F. D. Dalsenter. 2004. "A review of recent developments in modeling of microbial growth kinetics and intraparticle phenomena in solid-state fermentation." *Biochemical Engineering Journal* 17 (1):15-26.

Montoya, S. 2012. "*Obtención de enzimas lignocelulolíticas y polisacáridos a partir de residuos lignocelulósicos del Departamento de Caldas empleando macromicetos de pudrición blanca por fermentación sumergida y fermentación en estado sólido*." PhD Diss., Facultad de Ciencias Agropecuarias, Universidad de Caldas. ["*Production of lignocellulolytic enzymes and polysaccharides from lignocellulosic waste from Department of Caldas employing white rot macromycetes by submerged fermentation and solid-state fermentation.*" PhD Diss., Faculty of Agricultural Sciences, Universidad de Caldas].

Montoya, S., C. E. Orrego, and L. Levin. 2012. "Growth, fruiting and lignocellulolytic enzyme production by the edible mushroom *Grifola frondosa* (maitake)." *World Journal of Microbiology and Biotechnology* 28:1533-41.

Montoya, S., O. Sánchez, and L. Levin. 2014. "Evaluación de actividades endoglucanasa, exoglucanasa, lacasa y lignina peroxidasa en diez hongos de pudrición blanca." *Biotecnología en el Sector Agropecuario y Agroindustrial* 12 (2):115-24. ["Evaluation of endoglucanase, exoglucanase, lacasse, and lignin peroxidase activities on ten white-rot fungi." *Biotechnology in the Agricultural and Agro-industrial Sector* 12(2):115-124].

Montoya, S., Ó. J. Sánchez, and L. F. Gutiérrez. 2017. "Bioreactor to obtain bioactive substances through solid-state fermentation using macromycetes fungi." United States patent 9556409.

Moore, D. 1998. *Fungal Morphogenesis*. New York: Cambrige University Press.

Nelson, N. 1944. "A photometric adaptation of the Somogyi method for the determination of glucose." *Journal of Biological Chemistry* 153 (2):375-80.

Ohga, S., and D. J. Royse. 2001. "Transcriptional regulation of laccase and cellulase genes during growth and fruiting of *Lentinula edodes* on supplemented sawdust." *FEMS Microbiology Letters* 201 (1):111-15.

Papinutti, V. L., L. A. Dorio, and F. Forchiassin. 2003. "Degradación de madera de álamo por *Fomes sclerodermeus*: Producción de enzimas ligninolíticas en aserrín de álamo y cedro." *Revista Iberoamericana de Micología* 20:16-20. ["Degradation of poplar wood by *Fomes sclerodermeus*: Production of ligninolytic enzymes in poplar and cedar sawdust." *Iberoamerican Journal of Mycology* 20:16-20].

Pérez, J., and T. W. Jeffries. 1992. "Roles of manganese and organic acid chelators in regulating lignin degradation and biosynthesis of peroxidases by *Phanerochaete chrysosporium*." *Applied and Environmental Microbiology* 58 (8):2402-09.

Somogyi, M. 1945. "A new reagent for the determination of sugars." *Journal of Biological Chemistry* 160:61-68.

Suyanarayan, S., and K. Mazumdar. 1998. "Solid State Fermentation." US patent 6197573.

Thurston, C. F. 1994. "The structure and function of fungal laccases." *Microbiology* 140 (1):19-26.

Tuor, U., H. Wariishi, H. E. Schoemaker, and M. H. Gold. 1992. "Oxidation of phenolic arylglycerol β-aryl ether lignin model compounds by manganese peroxidase from *Phanerochaete chrysosporium*: Oxidative cleavage of an α-carbonyl model compound." *Biochemistry* 31 (21):4986-95.

Velázquez-Cedeño, M. A., G. Mata, and J. M. Savoie. 2002. "Waste reducing cultivation of *Pleurotus ostreatus* and *Pleurotus pulmonarius* on coffee pulp: changes in the production of some lignocellulolytic enzymes." *World Journal of Microbiology and Biotechnology* 18 (3):201-07.

Wong, Y. S., and X. Wang. 1991. "Degradation of tannins in spent coffee grounds by *Pleurotus sajor-caju*." *World Journal of Microbiology and Biotechnology* 7 (5):573-74.

Xing, Z. T., J. H. Cheng, Q. Tan, and Y. J. Pan. 2006. "Effect of nutritional parameters on laccase production by the culinary and medicinal mushroom, *Grifola frondosa*." *World Journal of Microbiology and Biotechnology* 22 (8):1215-24.

About the Authors

Sandra Montoya

Associate Professor, Department of Engineering. Universidad de Caldas
Director of the Bioprocess and Agro-industry Plant.
Universidad de Caldas, Colombia

Sandra Montoya is an Associate Professor at Universidad de Caldas, Colombia. She received B.Sc. and M.Sc. degrees in Chemical Engineering from the National University of Colombia, and a Ph.D. degree in Agricultural Sciences from Universidad de Caldas. She joined the Faculty of Engineering at Universidad de Caldas in 2004 where she has coordinated different undergraduate and postgraduate courses related to biotechnology and macromycetes. Her research focuses on biotechnology of macromycetes, composting, enzyme technology, and other agro-industrial bioprocesses. She has led research and development projects on the utilization of agro-industrial waste and macromycetes cultivation. Dr. Montoya has authored over 25 research papers and 2 book chapters, and presented more than 30 works at Colombian and international scientific events. She was granted a US patent on a novel bioreactor for solid-state fermentations. She is one of the main Latin American researchers in the cultivation of Grifola frondosa in tropical weathers and has improved the productivity of Ganoderma lucidum production under Colombian conditions. She is founder and Director of the Bioprocess and Agro-industry Plant at Universidad de Caldas, a center for research and technological development. Dr. Montoya has designed, started-up and managed different commercial facilities for production and transformation of macromycetes fungi in Colombia.

Óscar J. Sánchez

Associate Professor, Department of Engineering. Universidad de Caldas
Director of the Research Group on Food and Agro-industry.
Universidad de Caldas, Colombia

Óscar J. Sánchez is an Associate Professor at Universidad de Caldas, Colombia. He received a B.Sc. in Chemical Engineering and M.Sc. in Biotechnology degrees with the highest honors from the Moscow State Academy of Fine Chemical Technology (Russia), and a Ph.D. degree from the National University of Colombia. He joined the Faculty of Engineering at Universidad de Caldas in 1997 where he has coordinated different undergraduate and postgraduate courses related to biotechnology. His research focuses on bioprocess systems engineering, biofuels production, industrial microbiology, and enzyme technology. He has led research projects on the utilization of agro-industrial waste by biotechnological methods. Dr. Sánchez has authored over 37 research papers, 3 research books and 11 book chapters, and presented more than 40 works at Colombian and international scientific events. He was granted a US patent on a novel bioreactor for solid-state fermentations. He was Honorary Research Fellow in the Department of Chemical Engineering of University College London in 2006. From 2009, he leads the research group on Foods and Agro-industry at Universidad de Caldas and works at the Bioprocess and Agro-industry Plant as an Associate Researcher. Currently, Dr. Sánchez is the Director of the Ph.D. program in Engineering at Universidad de Caldas.

Diego F. Arias

Supervisor of the Bioprocess and Agro-industry Plant
Universidad de Caldas, Colombia

Diego F. Arias received a B.Sc. degree in Agronomy from Universidad de Caldas, Colombia, and currently is student of the M.Sc. program in Systems of Agricultural Production at that university. From 2007, he has worked in the cultivation of macromycetes fungi. From 2012, he is the supervisor of the Bioprocess and Agro-industry Plant at Universidad de Caldas where he plans and executes the procedures for production of spawn and inoculated substrates of different macromycetes species for their commercialization at different production scales. His research and production activities focus on the cultivation of Pleurotus ostreatus, Pleurotus eryngii, Lentinus edodes, Ganoderma lucidum, and Grifola frondosa. Agronomist Arias has authored over 5 research papers and his M.Sc. work is related to the cultivation of G. lucidum. He has also worked as a consultant for several macromycetes cultures at different production scales.

Index

A

B

C

D

G

H

L

M

N

O

P

Q

R

S

T

U

V

W

X

Y

Z

α

β